comets, meteorites and men

comets, meteorites and men

PETER LANCASTER BROWN

New York | TAPLINGER PUBLISHING COMPANY

To the comet hunters,
past, present and future

First published in the United States in 1974 by
TAPLINGER PUBLISHING CO., INC.
New York, New York

Published simultaneously in the Dominion of Canada by
Burns & MacEachern, Ltd., Toronto

Library of Congress Catalog Card Number: 73-16633

ISBN 0-8008-1734-6

CONTENTS

FIGURES IN THE TEXT

TABLES *page*

BLACK AND WHITE PLATES

ACKNOWLEDGEMENTS

Tokyo Observatory, plate 11; Royal Astronomical Society, plates 2, 8, 9, 12, 16–17, 51; American Meteorite Laboratory, plates 36, 41–2, 44–6, 49; J.L. White, plate 25; British Museum, plates 37, 39, 47; Australian News and Information Bureau, plate 40; Karl-Schwarzschild Observatory, plate 6; Mt Wilson and Palomar Observatories, Jacket Picture (Comet Humason 1962 VIII).

The author also wishes to acknowledge the generous assistance from his wife, Johanne, who has shared in all the routine drudgery.

INTRODUCTION

In former times the appearances of comets played a significant role in astrology: they were portenders of war, plague, famine and the deaths of notables. Nowadays they are recognized as rather unique astronomical bodies that may act as natural space barometers and provide reliable indications of the day to day strength of the Sun's particle emission which is an all-important factor in relation to the safety of astronauts on space missions. The streams of emitted solar particles – known collectively as the solar wind – interact with material in the cometary head and produce tails which from our vantage point on Earth often appear to stretch in brilliant fashion right across the night sky.

It may be as well to draw attention at this stage to some mistaken notions concerning the apparent movements of comets. Comets and meteoroids ('shooting stars') are often confused. Even in serious literature many authors are guilty of committing a howler in the use of the simile: "It shot across the sky as swift and brilliant as a comet".

A meteoroid will most certainly appear to shoot swiftly across the heavens, but not so a comet. A comet is situated far beyond the atmosphere and although it is in orbit round the Sun and may perhaps be travelling at speeds in excess of 300 kilometres per second at the time of perihelion, its naked-eye movement adjacent to the backcloth of stars will only be apparent from night to night observation.

A bright comet usually consists of four separate parts: the cloud-like head, called the coma; the star-like centre, called the nucleus; the tail; and a surrounding highly tenuous cloud of hydrogen. In many faint telescopic comets, however, the tail and star-like nucleus are absent.

Until the era of the space-age, probably no other transient celestial object has attracted the attention of the public at large more so than the sudden and often dramatic appearance of a bright comet with a long spectacular tail. The appearance of such an object is, nevertheless, a comparatively rare event, and the majority of comets are never seen by the naked eye and require long exposures with large photographic telescopes to render them visible. The average comet has little or no tail and is seen as a faint, spherical, nebulous cloud.

Since man first began observing the heavens, he has puzzled over the sudden appearances of bright comets which are unlike any other celestial objects. Even today there are many unsolved problems about them. The very definition of what constitutes a comet is highly controversial.

The word comet is derived from the Greek *kometes*, meaning literally: 'the hairy one', but it was probably an Egyptian who coined the description: 'a hairy star'. The ancient Chinese likened them to brooms – hence the transient 'broom-stars' of their chronicles. Recently a comet was defined in the following terms: A comet is a cloud of meteors, surrounded (at times) by a nebulous envelope, the whole system moving round the Sun in an orbital path having the form of a conic section similar to the planets. But to arrive at even this highly simplified definition of a comet has involved several hundred years of continuous observation and interpretation.

A comet has also been described more prosaically as "a great big bag full of nothing". This is of course a tongue-in-cheek exaggeration, but there is an element of truth in it. The head of a comet occupies more volume than the Earth, and sometimes it may be even larger than the Sun. Yet in spite of this relatively large size it contains very little solid material, and its total mass is extremely small. Most of the volume of the comet is empty space with a little gas and dust distributed fairly evenly throughout. To gain some idea of the average density of a comet one may use the analogy, say, of a volume of space measuring one cubic kilometre inside which 12 small marbles are arranged equidistantly.

Some of the questions yet unanswered about these enigmatic objects are concerned with their physico-chemical make-up and how they fit into the cosmogony of the solar system. Until the late 1950s the study of comets was a somewhat neglected backwater of classical astronomy, but the space-age has brought comets into the limelight along with asteroids, meteorites and meteors with which they may be associated sometime during their evolution. Comets occasionally pass very close by the Earth and indeed they may some times collide with it. They are definitely connected in some way with 'shooting stars', and when we think in terms of bodies larger than small meteorites, comets are the most numerous class of object in the solar system and may possibly be the most numerous objects in the entire universe.

The fall of a large meteorite with its associated brilliant fireball and frightening noises along with the occasional catastrophic effects when it hits the Earth, has also attracted much historical speculation. Nevertheless, these bodies have only been scientifically recognized for just over a century and a half. In spite of all accumulated evidence, scientists of the eighteenth century were reluctant to admit to their very existence. There is the well-known story of Thomas Jefferson, then President of the United States and also a respected scientist, who on hearing the news of the large meteorite fall in Connecticut in 1807 exclaimed: "I could more easily believe that two Yankee professors would lie than that stone would fall from the sky". Even today meteorites and fireballs are often confused by lay opinion with so-called visible thunderbolts, which owe their origin to electro-meteorological disturbances in the atmosphere of the Earth.

Meteorites are solid chunks of cosmic material which reach the surface of the Earth. They are divided into three principal types: stones, irons and stony-irons. The old familiar meteor, with its colloquial description 'shooting star' or 'falling star', is nowadays called a meteoroid. The term meteor is now often used as a general name to include all cosmic material which passes through the Earth's atmosphere whether it reaches the surface of the Earth or not.

Meteorites have long been utilized by races in earlier times. The Eskimo used meteoric iron extensively for spears, axeheads, knives and miscellaneous tools. In North America, fragments of meteorites have been found in burial mounds of the Hopewell Indians – a race which inhabited the region in prehistoric times. To these Indians, meteorites afforded a major source of iron. Chemical analysis has actually identified the various burial-mound, nickel-iron fragments with specific meteorites. One example showed that it came from a meteorite that fell at Brenham in Kansas and which later ended up over 1,600 kilometres (1000 miles) away – a sign of the remarkable length of the prehistoric Indian trading network.

Micrometeorite and micrometeoroid are names given to extremely small particles of cosmic matter – the latter term is usually restricted to descriptions of particles reaching the surface of the Earth having suffered no surface ablation during their passage through the atmosphere. Fireballs or bolides are meteors brighter than magnitude –4, about as bright as the appearance of the planet Venus to the naked eye.

Until the lunar rock samples were brought back by the US and Soviet moonshots and since the probes to Mars, recovered meteorites were the only sources of extraterrestrial material available for direct study. The rare class of stony meteorites known as carbonaceous chondrites have long been suspected of containing 'organized' organic elements, and the evolution of life on Earth has been attributed by some to its arrival in primitive forms via these meteorites.

During the twentieth century, the new science of meteoritics has been developed in order to study cosmic material. This new science is a potpourri of such physical sciences as geology, petrology, mineralogy, chemistry, physics, astrophysics and biochemistry. New laboratory techniques developed during the past two decades to study meteorites have led to consistent estimates about their ages, and this in turn has provided positive clues in the determination of the chronology and make-up of the Earth, the solar system, the Milky Way and the Universe itself.

CHAPTER I

Comets in History

According to Diodorus Siculus, writing about 44 BC, the Chaldeans and the Egyptians considered that comets were of great importance in predicting events. Apart from this reference, however, it is not clear whether these early astronomers were convinced that comets were celestial objects in the sense of the planets and the stars, since there was probably some confusion of comets with atmospheric events such as the aurorae. It is not known on what evidence Diodorus Siculus based his remarks, for neither Babylonian nor Egyptian inscriptions and texts so far examined make any direct reference to a comet.* It is only through the Greek and Roman observers that we can glean – via later Byzantine codices – very doubtful third-hand hearsay information.

In the Bible there are only vague references to early comets, and probably the most extant one is in 1 Chron. 21 : 16.†

Certainly in the Greco-Roman world comets were very influential objects, and classical authors wrote about them at great length. They were always precursors of fatal events and

* Recently the author has tentatively identified the Egyptian hieroglyph for a comet with a nameless hieroglyph ideogram which many years ago Sir Wallis Budge vaguely interpreted "woman with dishevelled hair". This hieroglyph is almost a replica of the Sky goddess Nut – except for the addition of the long-flowing hair tresses. According to the Greek writers it was the Egyptians who first alluded to the analogy of long females' tresses in connection with the appearance of a spectacular comet.

† "And David lifted up his eyes, and saw the angel of the Lord stand between the Earth and the Heaven, having a drawn sword in his hand stretched out over Jerusalem . . ."

were even attributed with powers to influence events directly. The births and deaths of emperors were alleged to be heralded by brilliant comet apparitions. The bright comet which appeared in 43 BC was supposed to be the soul of Julius Caesar transported to the heavens.

Democritus strongly believed that all comets were the souls of the famous who, having lived on Earth, were at death borne into the heavens as brilliant lights. Aristotle pronounced with all his weighty authority that comets certainly could not be true astronomical bodies and assigned them to inhabiting the upper regions of his three regions of air.

However, not all the philosophers of the Old World followed Aristotle's ideas. Diogenes, Hippocrates and several of the Pythagorean school believed that comets – like the planets – were wanderers among the stars. But the first serious challenger to the Aristotelian ideas was Seneca, Nero's tutor. By carefully observing the movements of contemporary Roman comets, Seneca, in his book *Questiones Naturales*, came to the conclusion that only in appearance did they differ from the planets and they moved in the same kind of paths in the sky.

Pliny, in his *Historia Naturalis* (AD 77), writes several passages about comets and the terrible significance of their appearances. "A comet," he observes, "is a particularly frightful body, and not easily atoned, it is usually a very fearful star and announces no small effusion of blood."

In the early years of Christendom the dark and sinister spirit that prevailed in the Middle Ages did not lessen the morbid beliefs associated with the sudden appearance of a bright comet. The comets of AD 451 and AD 453 supposedly announced the death of Attila; the comet of AD 455 that of the Emperor Valentinian; and every bright comet which appeared during the early medieval period, the Middle Ages, and even the Renaissance had itself affixed to the death or misfortune of a prominent historical figure. These beliefs were so widespread that (according to Pingré) the chronicles recorded in good faith comets which were never actually seen – such as the one of AD 814 which allegedly heralded the death of Charlemagne.

A drawing of the 'comet' or 'prodegy' of the year AD 1000 shows that under the apparent stimulation of terror very little objective scientific observation was actually attempted. Although the drawing is supposedly contemporary, it certainly does not date from the precise epoch but is, nevertheless, probably highly representative of the ideas of the time. Another bizarre drawing and observation is by Ambroise Paré the famous French surgeon of the sixteenth century, showing the comet of 1528. Paré described it in these terms:

> This comet was so horrible and so frightful and it produced such great terror in the vulgar that some died of fear and others fell sick. It appeared to be of excessive length and was the colour of blood. At the summit of it was seen the figure of a bent arm, holding in its hand a great sword, as if about to strike. On both sides of the rays of this comet were seen a great number of axes, knives and blood coloured swords among which were a great number of hideous human faces with beards and bristling hair.

When, what is now known as Halley's Comet appeared in 1456, the Turks with Mohammed II at their head, were besieging Belgrade which was being defended by Huniades, and according to one story both armies were seized with fear. Pope Calixtus III, himself struck with general terror, ordered public prayers to be offered up for deliverance from the comet and the enemies of Christianity.* When Halley's Comet appeared in April 1066, it was considered a precursor of the Norman conquest; according to a proverb at the time a new star meant a new sovereign. The *Anglo-Saxon Chronicle* reads:

> . . . In this year King Harald came from York to Westminster at Easter, which was after the mid winter in which the king (Edward the Confessor) died. Then was seen over all England such a sign in the heavens as no man ever before saw; some say it was the star Cometa, which some men call the haired star; and it first appeared on the eve of Litani-major, the 8th of the Kalends of May [24th April], and so shone all the seven nights.

In the New World comets were also treated as presages. A comet which appeared some years before Cortez conquered the Aztecs – either one in 1499, or more probably a previous one in 1490 – was reckoned as a positive sign that Quetzal-

* One story relates that the Pope actually excommunicated the comet, but this version has been shown to be an eighteenth-century hoax and put about by a French author out of humour with the church!

coatl would eventually return to Mexico. When Cortez and his knights arrived in 1519, this was confirmation that the great rain-spirit god from across the ocean had returned, and Montezuma accepted him without question.* Further to the north, comets were known to the Red Indians as spirits of the stars.

During the thirteenth and fourteenth centuries many of the spectacular comets which appeared are recorded in detail and often at great length. An anonymous treatise dated about 1238 commented on the fact that its author had noticed (correctly) that *bright* comets rarely appear except near the Sun.

Thomas Aquinas also recorded his cometary views which not surprisingly leaned heavily toward Aristotle's ideas. But in the sixteen centuries which followed Seneca there is practically nothing to tell about the advance of cometary science, and there are only the dreary monotonous – often hysterical – chronicles concerning astrological predictions and forebodings.

Fortunately the Renaissance, which had been so favourable to the arts, eventually began to influence scientific method. Peter Apian in 1531 astutely observed that comet tails always appear to be turned in a direction away from the Sun (see Figure 1), but the Chinese had recorded this fact about AD 800. When Halley's Comet appeared in AD 837, the Europeans were scared out of their wits while the more civilized Chinese were observing it with cool scientific detachment.

The turning point in the history of cometary astronomy came when Tycho Brahe, fired with zeal to make exacting observations of the planets and stars, applied himself to observing the bright comet which appeared in 1577. His own numerous observations, backed by simultaneous observations by a student friend in Nuremberg, proved beyond all reason-

* It seems very likely that the white and bearded god who appeared in the east associated with the Quetzalcoatl (Serpent God) legends of pre-Columbian Middle America relates to the apparitions of spectacular comets in the morning sky and *not* to the planet Venus which some suppose. The pre-Columbian Middle America races were a good deal interested in comets, which were symbolized by the plumed serpent depicted in various forms. Like the Chinese, the Middle America pre-Columbian races certainly made scientific observations of comets, but owing to the total destruction of whole libraries of their codices by the Conquistador priests, none are now extant.

able doubt that comets were bodies which moved in the region beyond the Moon and so confirmed a tentative idea put for-

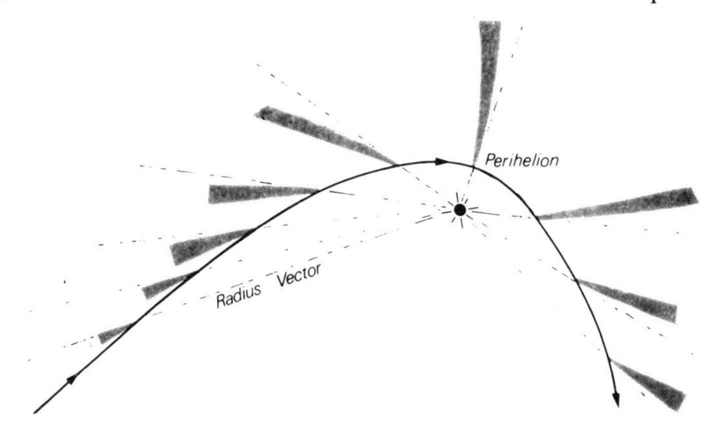

Figure 1. Comet tails, especially the spectacular 'dust' tails, always point away from the Sun, but they generally lag behind the Comet-Sun line (or radius vector).

ward by the famous Dr Cardano some years earlier. This also confirmed the inspired guesses of Seneca and some of the earlier Greeks that comets were true astronomical bodies. Tycho Brahe endeavoured to represent the movement of the comet by making it describe an orbit round the Sun external to Venus and vaguely specified its physical nature as having been 'engendered' in the depths of space.

Kepler, Tycho Brahe's protégé, investigated the movements of the comets of 1607 and 1618. He deduced from this work that comets traversed the solar system in rectilinear orbits – or in other words almost in straight lines. He also guessed (rightly) that comets were probably "as numerous in the heavens as are fishes in the sea".

A climax of scientific interest in comets occurred in 1618, when three bright ones appeared in rapid succession – the last of which was of unusual brilliance and remained visible over the period November 1618 until January 1619. Subsequently a plethora of discourses, pamphlets and books were inspired by this unprecedented celestial threat, and the debate concerning their true nature sparked off a great controversy between

Horatio Grassi (a Jesuit priest), Galileo, Mario Guiducci (a pupil of Galileo) and later Kepler. This is among the most interesting historical scientific controversies ever to take place.

Superstition was now on the wane. Bayle, in his essay *Pensées sur la Comète* wrote that the cynical wits of the seventeenth century eventually cast scorn and ridicule on the long cherished superstitions. He records the following story gleaned from the correspondence of a M de Bassompierre to M de Luynes in 1621, shortly after the death of Philip III: "... It seems to me that the comet we laughed about at St Germain is no laughing matter, as it has buried in two months a pope, a grand duke, and a king of Spain!"

It was Kepler's work on the paths of the planets which led to his discovery of the three well-known laws (see Appendix 1) which in turn led to the later solution of cometary movements. These laws revolutionized celestial mechanics, and he probably only omitted including cometary motion since a comet could not be continuously observed to make a complete path round the Sun as could the planets in their elliptical motions.

Sir William Löwer*, the great English dilettante scientist of the seventeenth century, when commenting on Kepler's work in 1610, suggested that the "unknown walkes of comets" may also move them in elliptical orbits. But more than half a century was to pass before final proof of cometary motion was forthcoming.

The Englishman Seth Ward put forward the idea of a circular or elliptical orbit for the comet of 1652; and in 1665 the French astronomer Auzout showed that the comet which had appeared in 1664 seemed to move in a straight line, and its path lay in one plane.

Hevelius in 1668, in his book *Cométographie*, vaguely suggested a parabolic orbit without substantiating the idea. In 1680, Doerfel, an obscure Protestant clergyman from Saxony and a pupil of Hevelius, published a small book, now extremely rare, consisting of five leaves in quarto and one woodcut on which are represented parabolic orbits. In this book he put forward the idea that comets must move round the Sun in

* Who distinguished himself as one of the commanders of the British fleet which opposed the Spanish Armada and who, in 1606, was imprisoned in the Tower by the court of the Star Chamber.

parabolic paths, and therefore the Sun itself must occupy one focus of the conic section.

When Newton returned to Cambridge after the plague in 1667, he had began to reflect deeply about gravitation, and in the years which followed, the problem was never far from his mind. In 1677 he was set thinking on a new tack when Robert Hooke (who was also very interested in the gravity problem) wrote to Newton suggesting that a projectile would describe an ellipse if the Earth's gravity varied inversely as the square of the distance, and that Kepler's third law might be attributed to the force of gravity acting between the planets and the Sun. Newton at first considered Hooke to be mistaken in this assumption, as he himself thought such a projectile would follow a spiral curve. But on further reflection and after some experimental figure work, he realized that Hooke was correct. This was later to have repercussions, for when Newton finally solved the problem, Hooke smartly stepped in to claim prior discovery. However, it is one thing to put forward a suggestion and another to construct a mathematical proof to substantiate it. There is no doubt that Hooke had greatly assisted Newton by putting forward his idea, but he did not have the mathematical originality of Newton to frame it in formal terms that could be put to the test. Later Hooke told Halley that he had discovered the fundamental laws some time before Newton, but had decided to conceal the fact lest it was not appreciated!

With Newton's formalisation of his ideas about gravity, the problem of cometary orbits was near to being solved, but it needed yet another genius to finally crack the age-long mystery.

CHAPTER II

The Story of Halley's Comet

The story of Halley's Comet has its beginning when Edmond Halley visited his friend Newton in Cambridge in August 1684, with the express purpose of finding out how Newton was progressing with the problem of applying the idea of gravity to astronomical bodies.

Edmond Halley* was born in London on 8th November 1656, and from an early age studied mathematics and astronomy. Since his father was a well-off salter, he was able to send young Edmond to Oxford, and in 1676, at the age of 20, he published his first serious scientific paper on the orbits of planets. In the same year he visited St Helena to observe the southern skies and to prepare a catalogue of the stars in this then unknown hemisphere. His father had provided him with an allowance of £300 a year in order that he could carry on with his work and at the same time be spared the pecuniary difficulties that beset many of his scientific contemporaries.

When he returned he was only 22 years of age, and he was hailed as a southern Tycho by Flamsteed, the first Astronomer Royal. His star catalogue, *Catalogus stellarum australium*, the earliest one to be produced with the aid of a telescope, immediately made him one of the most respected astronomers of his day.

On his historic visit to Cambridge, Halley was overjoyed to find that Newton had already proved the proposition that bodies gravitationally attracted would describe paths as

* Usually spelt Halley and pronounced Haley, but it is believed that Halley preferred the pronunciation Hawley.

ellipses. Halley urged Newton to publish his results, and it is possible that had it not been for the great influence of his friend, Newton – who by all accounts was an extremely shy and retiring man – might never have revealed his discoveries which had been uppermost in his thoughts for fully fifteen years before coming to mathematical fruition.

Despite the controversy with Hooke which followed the announcement, Newton reluctantly agreed to publish, and his manuscript was submitted to the Royal Society of London on 28th April 1686. However, the Society was short of money. The publication of the *History of Fishes* had gone far to deplete the funds of the Society, making it impossible to carry out another costly undertaking. The outcome was that Halley undertook sole responsibility and printed it at his own cost. It was said that Newton could not afford to underwrite the publication himself, but this is now known not to be true, and he could well have provided the means. When Newton's monumental *Principia* was finally published in 1687, it established him as one of the greatest scientific men of his day.

Newton had shown that comets obeyed the laws of gravitation just as the planets did. Although Newton's thoughts had occupied him for almost two decades prior to the publication of the *Principia*, it is surprising that in 1680, Newton is reported to have been politely incredulous when Flamsteed wrote to him suggesting that a comet which had disappeared into the Sun's rays and one that appeared on the other side were not two comets as many supposed, but one and the same. But this is difficult to understand, for as it has been said, Newton himself had long pondered about the problem and was fully aware that earlier Sir William Löwer had suggested that comets may move in elliptical (or parabolic) orbits. Very little intuitive reasoning would be required to show that a comet, owing to the geometrical orientation of its orbit in relation to the Earth and the Sun, would often seem to disappear into the Sun at perihelion and reappear again at a later date as it passed away. Five years later Newton had a change of mind, and in a letter to Flamsteed in 1685 he wrote that it seemed very probable that the comets of November and December in 1680 were the same one.

At a meeting of the Royal Society in April 1687, Newton's

third book *De Systemate Mundi* was presented to the meeting for discussion. The *History of the Royal Society* records:

> . . . It contained the whole system of celestial motions as well as the secondary and primary planets, with the theory of comets; which he illustrates by the example of the great comet of 1680–81, proving that [comet] which appeared in the morning in the month of Nov. proceeding, to have been the same comet, that was observed in Dec. and Jan. in the evening . . .

It was fitting that Halley, whose almost unbelievable generosity and enthusiasm had been so important a factor in the production of the *Principia*, should be the first one to reap fruit from its application. But the cometary harvest of Newton's findings took thirty years to ripen – for Halley was a man with many fingers in the scientific pie and he pursued many interests.* He became preoccupied with the great navigational problems of the eighteenth century, one of which was finding longitude at sea in an era before the chronometer had been perfected. There had been one possibility of Halley's starting his cometary work earlier when the Savilian Chair of Astronomy at Oxford fell vacant in 1691. We know that Halley was an eager candidate for the post in order to gain sufficient leisure time to pursue his long-shelved cometary research. But David Gregory was elected instead. A dispute with Flamsteed who referred to him as "a drunken sea captain" did not assist his application, and there can be no doubt that Halley's unorthodox religious views and quarter-deck language were regarded with alarm and influenced the decision.

He was secretary of the Royal Society between 1685 and 1693, and his many enthusiasms kept him preoccupied. From 1696 to 1698, he was deputy controller of the Royal Mint at Chester. Later he actually commanded a ship as a post captain which was commissioned by the Admiralty in 1698, charged with investigating the deviation of the compass with change of longitude. In 1702, he made two voyages to the Adriatic to advise the Emperor on the fortifications of Trieste, and in this work he followed directly in the traditions of Galileo who published a best-selling text-book on the subject. By this time his status was recognized, and when the Chair of

* Including the compilation of the first mortality tables.

Geometry became vacant at Oxford in 1704, he was elected without dissent and was now able to enjoy the academic freedom which he had so long sought after to tackle the comet problem ever lurking in the background of his thoughts.

Halley had always been particularly interested in the great comet which appeared in 1682. The positions for this comet had been obtained with great care, and he decided to pay particular attention to its path in the sky and try and fit it to a parabolic orbit. He also began calculating parabolic orbits for 24 other well-observed apparitions of comets with a view to publishing the result in a work he had long contemplated called *A Synopsis of Comets*. Printed in Latin, as was contemporary custom, *Synopsis Astronomiae Cometicae* fills less than 18 small quarto pages using fairly large type. The historical introduction is accomplished in a mere three pages, and the kernel of the whole work is a table giving the parabolic elements of 24 comets. The computing work executed in a pre-machine age was extremely tedious. Newton had provided the methods, and Halley accomplished the heroic work of the application and calculation. The joint amalgam of their genius and effort bore new scientific fruit, for when he had finished, Halley was excited to find that instead of the parabolic orbits he had expected, he was able to derive some decidedly elliptical ones. In particular the great Comet of 1682 was found to be moving in a plane only a little inclined to the ecliptic and was actually an ellipse of very great eccentricity. At its furthest extent from the Sun the comet retreated to the enormous distance of 5,475,000,000 kilometres (3,400,000,000 miles). Checking through the orbital characteristics of other comets on his list, he was immediately struck by the similarity of the 1682 comet with two others.

What puzzled Halley, however, was that if it was the same comet which had appeared in 1531, 1607 and 1682, why had it not returned after exactly equal intervals? His interest was now fully aroused. Checking back through all the available historical records, he soon came to the firm conclusion that there were many apparitions of previous brilliant comets which were most likely earlier apparitions of the 1682 Comet. Astronomers of the day were well aware of the precise clockwork regularity of the Earth's movement round the Sun, but

Table I **Parabolic Elements of 24 Comets** (after Halley)

Date of Perihelion Passage				*Ascending Node*	*Inclination*	*Motion*	*Perihelion Longitude*	*Perihelion Distance*
1337 June	2	6h	25m	84° 21′ 0″	32° 11′ 0″	R	37° 59′ 0″	0·40666
1472 Feb.	28	22	23	281 46 20	5 20 0	R	45 33 30	0·54273
1531 Aug.	24	21	18	49 25 0	17 56 0	R	301 39 0	0·56700 *
1532 Oct.	19	22	12	80 27 0	32 36 0	D	111 7 0	0·50910
1556 April	21	20	3	175 42 0	32 6 30	D	278 50 0	0·46390
1577 Oct.	26	18	45	25 52 0	74 32 45	R	129 22 0	0·18342
1580 Nov.	28	15	0	18 57 20	64 40 0	D	109 5 50	0·59628
1585 Sept.	27	19	20	37 42 20	6 4 0	D	8 51 0	1·09358
1590 Jan.	29	3	45	165 30 40	29 40 40	R	216 54 30	0·57661
1596 July	31	19	55	312 12 30	55 12 0	R	228 16 0	0·51293
1607 Oct.	16	3	50	50 21 0	17 2 0	R	302 16 0	0·58680 *
1618 Oct.	29	12	23	76 1 0	37 34 0	D	2 14 0	0·37975
1652 Nov.	2	15	40	88 10 0	79 28 0	D	28 18 40	0·84750
1661 Jan.	16	23	41	82 30 30	32 35 50	D	115 58 40	0·44851
1664 Nov.	24	11	52	81 14 0	21 18 30	R	130 41 25	1·02575
1665 April	14	5	15	228 2 0	76 5 0	R	71 54 30	0·10649
1672 Feb.	20	8	37	297 30 30	83 22 10	D	46 59 30	0·69739
1677 April	26	0	37	236 49 10	79 3 15	R	137 37 5	0·28059
1680 Dec.	8	0	6	272 2 0	60 56 0	D	262 39 30	0·00612
1682 Sept.	4	7	39	51 16 30	17 56 0	R	302 52 45	0·58328 *
1683 July	3	2	50	173 23 0	83 11 0	R	85 29 30	0·56020
1684 May	29	10	16	268 15 0	65 48 40	D	238 52 0	0·96015
1686 Sept.	6	14	33	350 34 40	31 21 40	D	77 0 30	0·32500
1698 Oct.	8	16	57	267 44 15	11 46 0	R	270 51 15	0·69129

inequalities had been noticed regarding the motions of Jupiter and Saturn. Intuitively Halley guessed that this was due to the gravitational attraction of the planets on one another in addition to the independent action of the Sun on each separate one. He reasoned that if the same situation were applied to comets, the inequalities of the two intervals between a comet's perihelion passage round the Sun might easily be accounted for, and he further argued that the effects might even be much greater in the case of a comet. Halley finally concluded that there were no difficulties which could not be explained away in such ways. As a consequence, he was able to calculate that the same comet would reappear near the Sun in the year 1758.

In spite of the general accolade afforded Halley and the ready acceptance by a majority of his contemporaries as to the correctness of his prediction, there were still a few lingering doubts. The previous observations of Peter Apian in 1531, and those by Kepler in 1607* were only approximate ones. Although the orbits found for the comets in the years 1531, 1607 and 1682 were strikingly similar, there could be two possible explanations: either three different comets followed the same path at nearly equal intervals of time (similar in some respects to the Sungrazer group of comets), or it was one comet describing an elongated elliptical orbit which differed little from a parabola in the neighbourhood of the Sun. Halley had guessed the perturbing effects of Jupiter fairly shrewdly, but he was also intuitively aware of perturbing forces of an unknown nature.

Halley died at the ripe age of 86 in 1742. When his *Astronomical Tables* (The Preface of the Synopsis) was reproduced seven years later in 1749 (second edition 1752, with English translation) it contained a firm reiteration from Halley – written some years before his death – that the comet would appear again in 1758.

Halley's prediction about a comet's return was certainly not the first one, for Jacques Bernouilli announced the return of the great comet of 1680 for 17th May 1719, and stated it would appear in the constellation of Libra, but no bright comet was seen.

* He also probably used those of William Löwer.

As 1758 approached, the orbit was recalculated in greater detail by Lalande, Clariaut and Madame Lepaute who are said to have toiled incessantly day and night – excepting meal times – on their monumental calculations. They also recalculated the perturbating effects of the planets, but they were handicapped by their ignorance of the two yet undiscovered massive planets, Uranus and Neptune. Scientific interest rose to fever pitch, and Voltaire said that the astronomers of France did not go to bed in 1758 for fear of missing the long-expected comet. But in spite of the monumental recalculation, the time of the reappearance was still uncertain. The known periods were quite unequal: from 1531 to 1607 the interval was 27,811 days, from 1607 to 1682 it was 27,325 days, which gives a difference of 459 days between perihelion passages round the Sun. The question remaining unanswered was, would the new period be shorter or longer – or would it be the old value?

In November 1758, Clariaut, who had devised novel methods to resolve this problem, finally presented the findings to the French Academy of Sciences. The concrete results of Clariaut's work was the prediction that perihelion passage would be delayed for 618 days, and that the actual year would be 1759. Saturn would delay the comet for 100 days and Jupiter 518 days, bringing the perihelion passage approximately to the middle of the month of April. Clariaut added that probably due to terms omitted (Uranus and Neptune discovered 1781 and 1846 respectively) and possible errors in calculation, there might be a difference of one month either side of the predicted date.

The great French comet hunter of the period, Messier, began searching for it in the middle of 1758. However, prior discovery went to a self-taught amateur astronomer, a Saxon peasant-farmer called Palitzsch, who with his home-made 7-foot reflecting telescope observed it on 25th December 1758. Subsequent observation soon showed that actual perihelion passage would take place on 13th March 1759, just 32 days before the time calculated by Clariaut. This work had now laid the foundation stone of modern classical celestial mechanics. Clariaut had shown that over an interval of 150 years of orbital motion, man could predict to within 32 days the position of a particular astronomical body in space.

The comet was due again in 1835. Spurred on by their predecessors, two Frenchmen, Damoiseau and Pontécoulant, independently undertook the equally laborious task of determining the epoch of the perihelion passage of the comet, taking into account the massive planet Uranus, discovered by Herschel in 1781. According to Damoiseau's reckoning, the comet should pass perihelion on 4th November 1835, and according to Pontécoulant not until 13th November. Two other astronomers who had entered the unofficial competition, Lehmann and Rosenberger, fixed the dates 11th November and 26th November respectively.

As early as December 1834, astronomers once more began their meticulous sweeping of the sky along the comet's orbital path which lay in the constellation of Auriga and Taurus – favourably situated for observation at this time of year. Nevertheless, the early sweepers were unsuccessful. The first glimpse of the comet was obtained on the morning of 6th August 1835 by Father Dumouchel and the astronomers of the Collegia Romano who had the advantage of the transparent Italian skies and one of the largest refracting telescopes in Europe. It appeared as a faint misty object hardly discernible with even the best optical equipment. On 23rd September, it was first seen with the naked eye, and the tail was noticed for the first time on the 25th, but this was not yet bright enough to attract the attention of the public until the end of the month.

A quick recalculation after the earliest observations showed that perihelion passage would take place on 16th November, only three days different to the prediction of Pontécoulant. This was an increase of 69 days over the preceding period, giving a true period of 76 years 135 days. Celestial mechanics had now reached new perfections of method, and Pontécoulant – spurred on by this success – immediately set about the task of calculating the next return, finally coming up with the prediction that the next perihelion passage would take place at 11.0 pm on 24th May 1910.

The 1835 apparition was particularly suitable for telescopic observation, and in October of that year the tail grew until it was fully 30° long. Observations continued until 22nd

November when the comet was hidden by the glare of the Sun. On 30th December, it was again picked up in more southerly latitudes as it emerged from the glare. Sir John Herschel, who at that time was at the Cape of Good Hope surveying the southern skies with a great 20-foot reflector, produced a long series of observations about the remarkable activity in the comet's head. (See page 62)

This apparition of Halley's Comet was an exceedingly bright one and well observed in both hemispheres. But there was little concrete evidence forthcoming to resolve the problem of the physical nature of cometary bodies. When the comet passed into the far southern skies, it gradually grew fainter as it receded from the Sun, and the last view of it was obtained by Sir John Herschel in May 1836 through his smaller 4-inch refracting telescope.

From time to time astronomers have attempted to retrace all the historical appearances of Halley's Comet. On first consideration this may seem a straightforward task simply by going back at intervals of 75 or 76 years. But it is not quite that easy. Halley was the first to encounter the difficulty. The comets of 1531, 1607 and 1682 gave him intervals of 76 and 75 years. Relying on this evidence alone a great comet should have appeared about the year 1456. That summer, when the Turks were threatening to overrun Europe, a bright comet was seen, and this is one of the most famous historical apparitions of Halley's Comet (see page 17). However, the mean period over two millenniums is not far short of 77 years, so that when Halley tried to go back to 1380 and 1305, he made errors of two and four years, and had he gone back further he would have been led even further astray.

The next positive identification was made by a French mathematician, Burckhardt (1773–1825), who investigated the orbit of a comet observed in 989 in China and also mentioned in several Anglo-Saxon chronicles. Three more appearances were identified by Laugier (1812–1872), an astronomer of the Paris Observatory. One of these occurred in AD 451, at the time of the celebrated battle of Chalons, when the Roman general Aëtius defeated Attila. Chinese records are sufficiently accurate to show that perihelion passage took place on 3rd July, and one is able to pinpoint the comet's path through the Pleiades into the constellations of Leo and Virgo – while by contrast the European 'observations' are mere babblings of hysterical nonsense.

Almost a century after the death of Halley, another Englishman appeared on the scene who was equally interested in this particular comet. J.R. Hind (1823–1895) in his own right discovered three comets and ten minor planets from the private observatory of Mr Bishop, which was then (in the middle of the nineteenth century) situated in Regent's Park, London. Hind is one of the few Englishmen to have gained prize money from his comet and minor planet discoveries. He became fascinated by comets at an early age, and apart from his observational work and the computation of the orbits of new comets he systematically examined old records, especially the Chinese Annals, to find out all he could about comets that had appeared in the past. From this research he was able to trace with fair probability many previous appearances of Halley's Comet as far back as 12–11 BC. At this apparition it terrified the Romans and appeared to them to be suspended directly over Rome; likewise it terrified the Jews in AD 66, when they were hard pressed by the Romans at the siege of Jerusalem. At this return, the comet, it is alleged, may also have been seen by St Peter, just before his martyrdom. Hind's work specified the 23 apparitions – relying heavily on Chinese observations made before AD 1400, since the European annals have little astronomical value apart from the 1066 apparition which is recorded on the Bayeux Tapestry (Plate 1).

This list remained definitive until the time of the next predicted apparition in 1910, when two British professional astronomers, Cowell and Crommelin, undertook, with the assistance of three helpers, Smart, Cripps and Wright, the mammoth task of proceeding backwards with each apparition step by step, making sure of each return before going on to the next. In this way they recomputed the predicted perihelion passage for 1910. They were able to extend Hind's list by finding two earlier returns, one in 87 BC and the other in 240 BC. They confirmed the 1066 apparition, but Hind was found to have been wrong in 608, in 912 and in 1223. (See Table II)

TABLE II

Returns to Perihelion of Halley's Comet

(Cowell and Crommelin)

240	(BC)	May 15?		912	(AD)	July 19	
163	(BC)	May 20?		989	(AD)	Sept. 2	(Burckhardt)
87	(BC)	Aug. 15		1066	(AD)	Mar. 25	(Hind)
12	(BC)	Oct. 8	(Hind)	1145	(AD)	April 19	(Hind)
66	(AD)	Jan. 26	(Hind)	1222	(AD)	Sept. 10	
141	(AD)	Mar. 25	(Hind)	1301	(AD)	Oct. 23	(Hind)
218	(AD)	April 6	(Hind)	1378	(AD)	Nov. 8	(Laugier)
295	(AD)	April 7	(Hind)	1456	(AD)	June 2	(Halley, Pingré)
374	(AD)	Feb. 13	(Hind)	1531	(AD)	Aug. 25	Halley
451	(AD)	July 3	(Laugier)	1607	(AD)	Oct. 26	Halley
530	(AD)	Nov. 15	(Hind)	1682	(AD)	Sept. 14	Halley
607	(AD)	Mar. 26		1759	(AD)	Mar. 12	
684	(AD)	Nov. 26	(Hind)	1835	(AD)	Nov. 15	
760	(AD)	June 10	(Laugier)	1910	(AD)	April 19	
837	(AD)	Feb. 25	(Hind)	1986	(AD)	Feb. ?	

Cowell and Crommelin finally derived the computed perihelion passage as occurring on 16·6 April 1910. As 1910 approached, it was the old story of all the comet observers eagerly scanning the heavens in hope of being the first to spot the comet's reappearance. Observational work began during the winter of 1908–9. For the first time in history the photographic patrol plate was used instead of the less sensitive human eye to detect the comet. At the Yerkes Observatory, Chicago, O.J. Lee exposed his earliest plate on 22nd December 1908. But first to announce discovery was Dr Max Wolf of Heidelberg who spotted a minute image on one of his patrol plates which had been exposed on 11th September 1909, within 10 minutes of arc of the position predicted by Cowell and Crommelin. At this time it lay at a distance of 512 million kilometres (331 million miles) from the Earth.

After the first announcement was made, other observers eagerly searched their early plates. The Royal Observatory, Greenwich, had recorded it on 9th September, and Helwan Observatory,* in the transparent Egyptian atmosphere, as early as 24th August, and even Wolf found that he had overlooked an earlier image of 28th August. The actual perihelion passage turned out to be correct to within three days, yet in some respects this was a somewhat disappointing result for Cowell and Crommelin. It was one which, nevertheless, outrivalled their computing competitors who assigned a date two months later.The Lindemann prize of 1000 marks which had been offered by the Astronomische Gesellschaft for the most successful prediction was afterwards duly awarded to Cowell and Crommelin.

This 3·03 day discrepancy puzzled Cowell and Crommelin, for later on 5th March 1910, they wrote:

> It now appears from observations that the predicted time 16·61 April is 3·03 days too early. At least two days of this error must be attributed to causes other than errors of calculation or errors in the adopted positions and masses of the planets . . .

Cowell and Crommelin had shown that the period of

* The guiding telescope of the photographic reflector at Helwan was then the selfsame small refractor of John Herschel in which he had last seen the comet at the Cape 73 years earlier.

1835–1910 was the shortest on record. The longest had been between AD 451–530 and was nearly five years longer, showing the great effect that planets have on perturbing the comets, in particular the significant mass of Jupiter.

In 1909, the comet was first seen non-photographically by Burnham on 15th September, while observing with the 40-inch refractor at Yerkes Observatory. Barnard, a colleague, was able to see the comet with the 40-inch on 17th September, which was one of his scheduled nights. The relationship between these two great observers was often strained, and they were frequently in competition with each other. There is an interesting anecdote which illustrates the keen rivalry between the two. The story goes that Burnham, after a night's spell at the 40-inch refractor while Barnard had been working with the 12-inch photographic reflector in a nearby dome, casually mentioned at breakfast that he had seen a comet in one of his double star fields during the previous night. Burnham – not particularly interested in comets – asked Barnard with some polite indifference which one it was. Barnard was horrified, since at the time there were no comets observable in the sky, and he pressed Burnham to tell him in which of the scheduled double star fields he had seen it. Burnham, occasionally a somewhat absent-minded man for detail, said he couldn't remember for sure, but offered to let Barnard see his list of doubles for the previous night. The story relates that Barnard, not wishing to let any new comet escape him, spent the next month searching for Burnham's elusive comet – but unfortunately without result. It is said that Barnard was so annoyed at missing the comet that he barely spoke civilly to Burnham for many months afterwards.

During September and the remaining months of 1909, Halley's Comet gradually brightened up as it approached the Sun. By 2nd February 1910, it was observed as a faintish nebula with small telescopes. In the latter part of March it passed out of view behind the Sun and did not reappear until the end of April when it was seen to have a splendid tail easily visible to the naked eye. On 6th May, it had reached magnitude 2 (2^{m}),* and the tail had divided into two parts. On 17th

* Magnitudes are expressed in shortened form, i.e. 5^{m}=magnitude 5. (See also Appendix VI).

May, the tail reached 70° in length and 9° in width. In the southern part of the northern hemisphere and in the southern hemisphere itself it was the most brilliant and dominant object in the sky.

On 18th May, the orbit of the comet carried it between the Earth and the Sun. Because of the unique event, an American expedition was specially sent out to Hawaii to observe the Sun's disc at the critical time, but there was not the slightest trace of the comet's nucleus projected on the background of the Sun's disc. It is concluded that had there been a solid body 320 kilometres (200 miles) in diameter forming the nucleus, it would have been readily spotted by observers.

On 19th May, reports showed that 75° of tail had been visible above the horizon,and there had been at least 30° of tail below—so that altogether it had reached a total length of 105°. On the day the Earth was due to pass through the comet's tail, a special lookout was kept to watch for any peculiar atmospheric/meteorological effects which could be attributed to the event. But the Earth only passed through the outer edge of the tail and was at least 8 million kilometres (5 million miles) from the comet's head. Owing to brilliant moonlight no positive atmospheric effects could be confirmed, but, as expected, the wide publicity afforded the event by the press gave rise to hundreds of eye-witness reports of alleged 'unusual' happenings in the sky.

At 2.30 am on 19th May, a soft glow could be seen diffusing from below the constellation of Cassiopeia. A few hours later the tail streamers were distinctly visible in spite of strong moonlight, and the 'train' of the comet stretched across the sky for at least 140°. However, the brilliant display was now coming to an end, and as the comet gradually drew away from the Sun, it began to fade rapidly. In early June, part of the tail which had broken away could be traced as it receded from the head of the comet at 60 kilometres per second. By 12th August, the comet had long been lost from naked-eye view and was about 9^{m}. It was last seen visually in a telescope at Helwan in Egypt on 29th April 1911, and was followed photographically until June by which it had receded from the Sun to a distance of 837,000,000 kilometres (520,000,000 miles).

In the 1835 apparition the comet had been visible for a total of 650 days. In 1910, it was discovered 249 days before perihelion passage and was observed for 396 days after perihelion – giving a total time of 645 days. Measurements showed that the head reached its greatest diameter of 550,000 kilometres (340,000 miles) on 8th June 1910, and it must have been greater than 970,000 kilometres (600,000 miles) if the observable halo surrounding it was included. At maximum the tail reached almost 48,000,000 kilometres (30,000,000 miles) – a little less than a third of the Earth's distance from the Sun. After June 1911, the comet finally disappeared and will not be seen again until it returns to the Sun in 1986.

Meanwhile, each year we are reminded of Halley's Comet by an annual meteor shower known as the Eta Aquarids – so named because they appear to radiate from a position near the star Eta (η) in the constellation of Aquarius. The meteoroid particles belonging to this meteor shower move in orbits very similar to that of the comet's orbit. There can be no doubt that there is some genetic connection between them and the comet, but the precise nature of the relationship between comets and meteoroids is not yet fully understood (see page 208). The Eta Aquarid meteoroids are spread out along the entire length of the comet's orbit and extend for 18,000,000 kilometres (11,000,000 miles) on either side of the comet's path so that it takes the Earth about two weeks in late April and early May to traverse completely through them during its journey round the Sun.

The problem of the orbit of Halley's Comet has not yet been completely solved, and it will be recalled that Cowell and Crommelin attributed discrepancies of over two days to causes other than errors of calculation. In recent years much attention has been paid to the yet unexplained so-called anomalous, non-gravitational forces which may be at work affecting the secular motions of comets.*

The 1986 return of Halley's Comet has received a great deal of attention from the contemporary orbit computers armed

* Recently an idea has been revived that the anomalous motion of Halley's Comet is due to the presence of a massive unknown planet lying beyond Pluto. However, this would still not account for the anomalous decelerations and accelerations apparently acting discriminatingly on a small select number of comets whose orbits lie inside Jupiter's orbit.

with the most powerful electronic tools that so far have been brought to bear on the problem. A date and time as precise as February $5^{d}.36775$ for perihelion passage has already been published by one of these computers. How correct this prediction is will only be known after the comet is resighted some time in 1985.

CHAPTER III

Comet Nomenclature, Orbits and Catalogues

Every year brings under observation a dozen or more comets. Some of these comets are new ones not previously recorded, but others are simply reappearances of well-known periodic comets, such as Halley's, which return to the Sun at precise or fairly precise intervals.

In order to keep track of all these comings and goings of comets, a system had to be evolved and drawn up with international agreement so that when a comet is first observed, whether it be a new one or the return of a known one, it is provisionally labelled and identified by a letter of the alphabet affixed behind the year of observation. In this way the first comets in 1971 for example are labelled 1971*a*, 1971*b*, 1971*c*, etc. in alphabetical order. A year or so later when their orbits have been thoroughly investigated, the comets are rearranged in order of date of perihelion passage and given a permanent Roman numeral instead of letters i.e. 1971 I, 1972 II, etc. These then become definitive designations which stand for all time and are entered into the comet catalogues in chronological order. This system allows for the later exclusion of some doubtful comets which although announced as discoveries, are never seen again and therefore not confirmed. In addition to the Roman numerals, a comet, if it is a new one, will also receive the name of its discoverer or discoverers. This sometimes leads to complications if a number of people independently discover the same comet as often happens in the case of a very bright one. To overcome this sometimes thorny problem it has been internationally agreed that no more than three names may be attached to any one comet. The first three to announce their independent discoveries are the ones to have their names attached.

The ultimate authority for looking after all the comings and goings of comets is the Telegraph Bureau of the International Astronomical Union (IAU for short) which is the recognized body responsible for monitoring all discoveries and interesting observations. At the present time the IAU Telegraph Bureau is located at the Smithsonian Institute at Cambridge, Massachusetts, USA, and its operation is run on a day-to-day basis by a full-time bureau director whose task is to keep abreast of astronomical events (a large proportion of which are cometary ones). When a new discovery claim is made, usually by telegram, the bureau director, after verification, initiates an IAU telegram which is dispatched to all IAU subscribers throughout the world. A few days later an airmail card is sent which contains more definitive details and perhaps additional observations which have been received by the bureau since the first announcement.

Usually when a periodic comet is reobserved, it does not acquire any new or additional names, but occasionally this rule may be broken. For example, Comet P/Perrine (the prefix letter P before a comet always denotes it is a periodic one), which was missed for six returns, was rediscovered quite accidentally by the Czech Anton Mrkos in 1955, and is now known by both names i.e. P/Perrine-Mrkos. A comet may also bear a name other than its discoverer, and Halley's Comet is perhaps the most famous example. Encke's Comet was first seen by the French comet hunter Méchain from Paris in 1766. It was observed again in 1795 by Caroline Herschel, and was discovered independently in 1805 by Pons from Marseilles, Huth at Frankfurt-on-Oder and Bouvard in Paris. It was observed accidentally again by Pons in 1818, and when the orbit computer Encke undertook a rigorous investigation, he was able to retrace its history back to Méchain's Comet of 1766. Consequent to Encke's work on the comet it has been called after him, although astronomers in the Soviet Union often refer to it as Comet Encke-Backlund, including the name of another computer who at a later date also investigated this comet's orbit.

In recent times a comet long known by the awkward title P/Pons-Coggia-Winnecke-Forbes – which is compounded from the surnames of comet hunters who independently rediscovered it at different historical apparitions – was renamed P/Crommelin nine years after the death of this British computer who investigated its orbit and, it may be remembered, shared the prize in 1910 with Cowell for the best prediction of the perihelion passage of Halley's Comet.

In order to compute the orbit of a comet a minimum of three observed positions is required, each observation defining at an instance in time its celestial coordinates Right Ascension and Declination – the analogues of longitude and latitude on the Earth's surface. When the first three observations are to hand, this is perhaps the most exciting time for the desk-bound astronomer/computers who otherwise lead rather routine lives and rarely actually observe a comet. There is an element of competitive spirit in this field similar in some respect to the competitive spirit among the comet hunters. Prior to the introduction of the modern electronic computing machines, the task of deducing an orbit would take many hours of tedious work using mechanical calculators. Nowadays, however, electronic computers can perform the task in a matter of minutes which is often less time than it takes to organize the results into the appropriate IAU telegraphic coding for transmission!

But not all new orbits are that easy to determine. Usually the first observations – especially the discovery observation – are only approximate ones, perhaps hasty and inaccurate estimates by an ebullient amateur comet hunter eager to stake his claim. If a comet is observed close to the Sun and in a twilight sky, precise positions are difficult to measure, and the computers will start biting their finger nails if their calculated ephemerides for the comet begin to deviate from the subsequent path of the comet.

Orbit computing can be one of the most fascinating exercises of the human mind, and even today much of the work is performed by self-taught amateur astronomers working at home armed with nothing more than portable hand calculating machines. The British Astronomical Association has in its ranks a band of amateur enthusiasts (with highly professional standards) who, with the help of financial grants from the IAU, have produced a magnificent catalogue of all known comets and their orbits which is acclaimed as an outstanding contribution to cometary astronomy everywhere in the world.

Comets differ considerably in their physical appearances, and any particular comet can in fact appear radically different at different apparitions. Without knowing the orbital characteristics of a comet it would be impossible to identify a periodic comet at different apparitions simply by its size, brightness or tail structure.

The comets which have enormously elongated orbits that stretch out beyond the orbits of the planets Neptune and Pluto – and beyond the known boundaries of the solar system – tend to be brighter objects than the comets with orbits stretching only as far as Jupiter's orbit. This difference is attributed to the idea that the short period comets have lost more material owing to their more frequent visits to the neighbourhood of the Sun. The solar wind particles interacting with the solids in the cometary head are likely to cause considerable erosion near the time of a comet's perihelion and so bring about a diminishment in the material which forms the head.

The duration of a comet's visibility varies from a few days to more than a year. Comet Schwassmann-Wachmann (1) has an almost circular orbit and can be seen each year at the time of opposition with the Earth. Comet Stearns (1927 IV), a huge comet which came to perihelion at the distance of Jupiter's orbit, was observed for four years. But the most usual period of visibility of a comet does not exceed two to three months.

The observed brightness of a comet not only depends on its physical characteristics but also on its geometrical orientation in relation to the Earth and Sun. Comets are not self-luminous bodies like the Sun and other stars. Part of their light is due to the reflection of sunlight from tneir solid of quasi-solid particles much in the same way as the planets reflect sunlight from their outer surfaces. They also owe part of their illumination to fluorescence which is a property which allows the gases in the comet head to absorb sunlight at one particular wavelength and then to re-emit the radiation at another wavelength. If the light of a comet is examined in a spectroscope, it usually shows the Fraunhofer lines of the dark absorption (or reflected light) spectrum and also the bright lines of the emission (fluorescence) spectrum superimposed.

Some comets can be observed at a considerable distance from the Sun. The remarkable Comet of 1729 never drew closer to the Sun than the radius of Jupiter's orbit, yet it was visible to the naked eye for nearly six months. Other comets, for inexplicable reasons, may be first seen as bright objects and then suddenly fade and become lost from view in even the largest telescopes.

If a comet is close to the Earth, it may appear to pass over a considerable angular distance of sky in a few days whereas the 1729 Comet – because of its great distance from the Earth – only traversed a length of 15° (or three times the distance between the Pointer Stars in the Great Bear) during the six months it was under naked-eye observation.

Kepler was the first to remark that comets may be as numerous as the fishes in the sea, and modern ideas agree that this was no idle speculation. About 300 new comets are identified every century, and many which approach the Sun and the Earth undoubtedly escape detection for various reasons. A total (solar system) population ranging from 100,000 to 150,000 comets is generally agreed upon among cometary astronomers. But this is really a minimum number, and figures as high as 5 million have often been put forward. This includes among its members comets whose orbits are such elongated ellipses that at aphelion they stretch out into space half-way towards the nearest stars. From our position on Earth we can only hope to see a tiny selection of comets. The Earth is 150 million kilometres (93 million miles) away from the Sun, and it will be recalled that the 1729 Comet never came closer than four times this distance. It is extremely likely that there may be a vast population of comets which come to perihelion at the distance of Jupiter and Saturn, but most, at this distance, would be too faint to be seen. There may also be a vast population of 'pygmy' comets which are too faint to be observed in present day telescopes – except those whose orbits carry them close to the Sun's surface (like the brighter Sungrazers) which have long been suspected on photographs taken at the time of total solar eclipse.

In the account of Halley's Comet it will be remembered that it was Newton who first showed that comets obeyed the gravitational principles formulated by him in the *Principia*. However, Newton's ideas had developed as a consequence of earlier work and in particular that of Johannes Kepler whose study of the orbit of Mars allowed him to arrive at his famous three laws concerning planetary motion.

The first of these laws shows that a planet moves in an ellipse with the Sun at one focus. The second law states that a straight line connecting the planet with the Sun (known as the radius vector of the planet) sweeps through equal areas in equal intervals of time (Figure 2). His third law came later and showed a definite correlation in the sizes of the planetary orbits and their orbital periods round the Sun.

Kepler had only been able to arrive at these results by

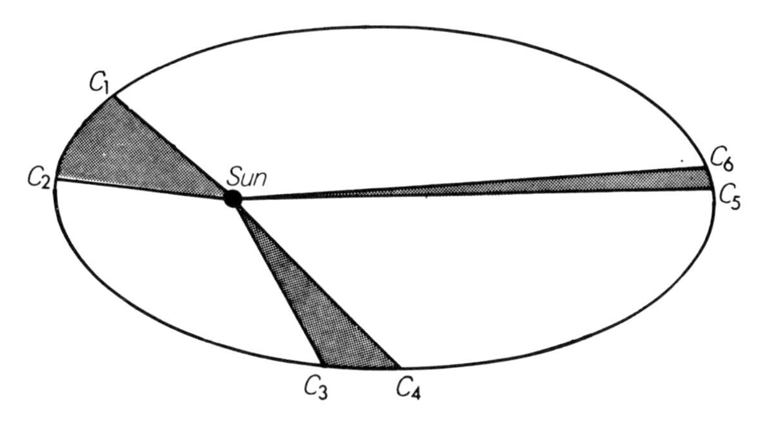

Figure 2. Kepler's second law (of equal areas). A comet (or any body in elliptical orbit) takes the same time in moving from C_1–C_2 and C_3–C_4, and C_5–C_6. The swept area C_1–S–C_2 = C_3–S–C_4 = C_5–S–C_6.

having access to Tycho Brahe's observations. It must be emphasized that without Brahe's painstaking practical observations, Kepler could not possibly have made the necessary deductions.

However, the story of planetary and cometary motions has its roots much further back in history. It was Apollonius of Perga (260–200 BC) who first experimented with conic sections and pondered briefly about the mathematical properties of an ellipse. Unfortunately this early work led directly into the blind alley of the idea of egg-shaped planetary paths and Ptolemy's highly involved epicyclic system which was later to take 1,300 years of scientific reasoning to eradicate before the Copernican system finally ousted it. Yet, ironically, the Copernican system itself was incorrect in so far as it adopted the ideas of circular motion for the planets.

It was not until Kepler's elliptic ideas were published that astronomy was finally eased of its Greek obsessions with the perfect symmetry of the circle and sphere. When Newton extended Kepler's work, he showed that any body moving in space by the Sun's attraction must move either in a circle, parabola, ellipse or hyperbola (Figure 3). In the case of the circle and ellipse the orbit closes back on itself while the parabola and hyperbola are open-ended curves.

The forces which dictate this curvilinear motion are the *cen-*

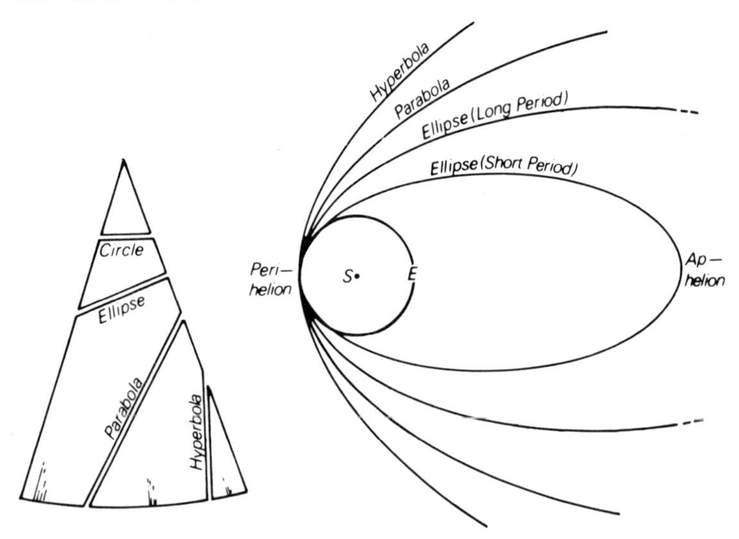

Figure 3. Conic sections and cometary orbits: (left) A cone can be sectioned to produce a circle, an ellipse, a parabola and a hyperbola. (right) Near perihelion the different cometary orbits are often indistinguishable.

tripetal – influencing a body inwards (i.e. gravity) – and the *centrifugal* – influencing the body outwards. These two forces counterbalance each other, and therefore it is the mutual tug-of-war relationship of these opposing forces which determines whether the orbital path (curve) described will be a circle, parabola, hyperbola or an ellipse.

With a body in orbit, the curved path induced by the two forces will produce a circle if the two forces are exactly equal. However, in practice, when the probabilities of the kind of curve described by a body in space are calculated, the chances of either a circle or parabola occurring is extremely small. In both these kinds of orbit the balance of the two opposing forces is exactly equal, or stated another way, it implies that a particular velocity is absolutely necessary, for even the slight increase or decrease of one component will cause a circular or parabolic orbit to degenerate into an ellipse or hyperbola. It is not surprising therefore that the orbits of the planets and their satellites are elliptical. Any body moving in a hyperbolic orbit would be non-periodic and on rounding the Sun it would then hurl away into deep space never to be seen again unless by sheer coincidence a nearby passing star perturbed it back towards the solar system.

Halley's investigation of cometary orbits showed that the so-called parabolic path of the 1682 Comet was indeed an elliptical one with a period of revolution round the Sun of approximately 75 years. From our vantage point on Earth, at one astronomical unit (1 AU) from the Sun, we can only see a fractional part of the total path of a comet, whereas with the planets, whose orbits are ellipses that are almost circular, we can observe them for long periods over their entire curvilinear motions round the Sun. Halley soon realized that one of the difficulties peculiar to comets – which has always beset the orbit computers – is on account of the actual shape of the apparent path taken by an elliptical (periodic) comet whose eccentricity is great so that it is very difficult to differentiate it from the shape of a parabola (Figure 3). Newton, the master, described the problem of orbit determination as "*Problema longe difficillium*", and it still taxes the geniuses of today armed with powerful electronic computers which are able to solve complex problems in celestial mechanics within millisecs.

In order to determine the orbit of a comet six elements are required (see also Figure 16):

1) The perihelion distance of the comet (or distance from the Sun at its closest approach.)
2) Its heliocentric longitude (a position in relation to the Sun, nowadays termed the argument of perihelion).
3) The position of the nodes (or when it crosses the ecliptic plane).
4) The inclination of the comet's orbit to the ecliptic plane.
5) The exact time of perihelion passage.
6) The period of the comet's revolution.

In practice one of the most difficult quantities to find with accuracy is the length of the major axis of the ellipse (Figure 4) which is the part that affects the determination of its period of revolution round the Sun. A small error of only a few seconds of arc can lead to differences of several hundred years in the calculation of the period!

In order to avoid this difficulty the procedure for computing

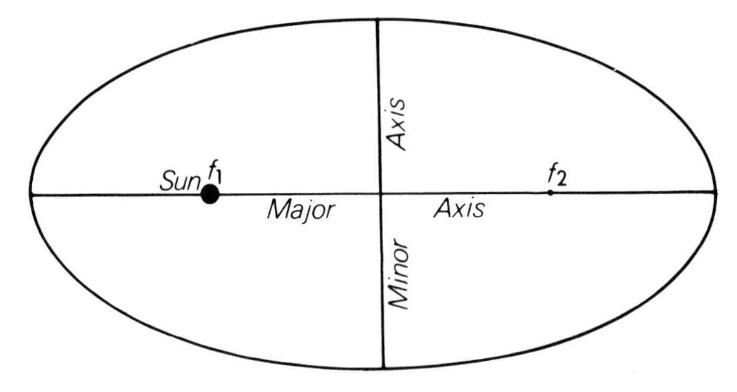

Figure 4. The properties of an ellipse. The eccentricity is measured by expressing the distance between the two foci f_1 (Sun) and f_2, in terms of the length of the major axis.

an orbit can be greatly simplified by leaving out the problem of a comet's revolutionary period by supposing that its orbit is that of a parabola, thus it would be supposed to be a conic section whose axis is infinite – or does not close back on itself. Of course this is really a form of mathematical 'cheating', for it has already been stated that a parabolic path is the least likely situation we would encounter in practice, but this assumption eases the computers' work immensely when computing the orbits of new comets.

When sufficient observations have accumulated over an extended observed orbital arc of the comet, the orbit can then be computed on a definitive basis. Some comets which have highly elongated elliptical paths, such as the Sungrazers, can never be given accurate orbital periods owing to similarity of their ellipses to a true parabola, and their orbital periods round the Sun may range between anything from 1,000 to 10,000 years or more.

The orbit of a comet may be computed in the simplest terms as a two-body problem using the first rudimentary method formulated by Newton. In this method it is assumed that both masses are spherical and homogenous. By a series of further assumptions and trial and error approximations a reasonably accurate result may then be achieved. To find the six orbital elements that define an orbit (see above) six* independent quantities must first be obtained either by direct visual observations or more accurately by measurement of the comet's image recorded on photographic plates. These are actually coordinate measurements of Right Ascension and Declination obtained by angular offsets to the adjacent field stars whose positions are recorded in various catalogues and known to a high degree of accuracy and then corrected for all the known variable factors.

When it comes to the problem of calculating a *definitive* orbit, the computer needs to take into account the gravitational attractions of all bodies which may influence the comet's path in the sky. This constitutes what is known as the *n* body problem, and the classic example of this kind is the situation where the planets pull against one another and are pulled by the Sun at the same time. Because of the differential motion of the planets as they revolve in orbit, the gravitational influences are in a constant state of flux. Although the Sun is the all-dominating factor owing to its colossal mass as compared with the majority of the planets, Jupiter's mass is also quite significant in perturbing the motion of a comet. When calculating the advance ephemeris of a predicted periodic comet, it is necessary to take all the known perturbing influences into

* Two in each observation.

account although in practice the *n* body problem is not strictly solvable because of its extreme complexity.

Any planet may perturb a much less bulky comet quite appreciably if it passes close by, and the comet's orbital energy may be increased or decreased substantially by such an encounter.

Unlike the major planets, many comets move in paths which are inclined to the ecliptic plane at steep angles. When the inclination exceeds 90°, they are said to possess retrograde motion round the Sun, which means that they revolve in a clockwise direction or the *opposite* (reverse) motion to that of the planets.

Comets as a complete family of bodies show a fairly even distribution of orbits about the sky. However, by statistically analysing various orbital characteristics such as perihelion distance, eccentricity, etc, it is apparent that some comets cluster in well-defined orbital groups. Comets may also be divided by their periods of revolution round the Sun. Comets with short periods go round the Sun in the interval of a few years with periods comparable to the planets. Comets of long period take several hundred, often several thousand years, to make one revolution. The distinction between short and long period comets has a hazy demarcation line, but one may say any comet with a period of *less* than 150-200 years is of short period.

A highly important major group of the short-period comets are those with periods comparable or less than the orbital period of Jupiter (11.86 years). This group is actually called the Jupiter family, since it has been demonstrated that most of these comets have at some time during the past come under its dominating influence and have been captured from long period orbits. Significantly all these Jupiter family comets have small inclinations to the ecliptic plane as would be expected of such bodies if they were to be effectively perturbed by the close encounter with a planet.

The orbital periods of the 40 or so of the short-period Jupiter family range from 3.3 to 8.6 years, but most of the periods lie between 6.3 and 7.8 years. There is a commensurability gap* in the orbital periods between 5.5 and 6.2 years which is very close to half the revolution period of Jupiter. Over an interval of time the comets switch from one side of the gap to the other. Only 5 of the 8 or more comets now populating the inner side of the gap have remained there since the beginning of the eighteenth century. The passage across the gap occurs in two stages. A close approach to the perturbing bulk of Jupiter sets the comet almost exactly in resonance, and 12 years later the next encounter pushes it across. Occasionally, however, the second approach to Jupiter switches the comet back to its original side of the gap. Although Jupiter's massive bulk plays a dominant role in the orbital evolution of comets, by no means all comets which come under its perturbing influence become family members of very short period.

Other giant (Jovian) planets such as Saturn, Uranus and

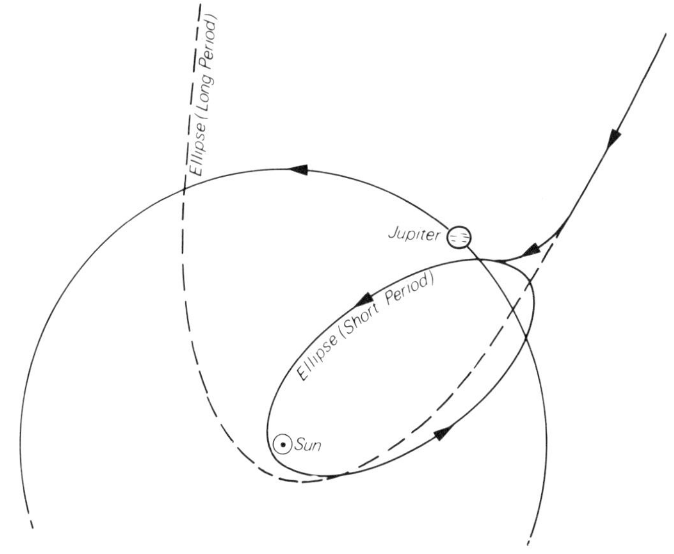

Figure 5. Capture by Jupiter of a long (indeterminate) period comet.

Neptune may also be influential in forming comet groups, but at present this is not considered proven. At one time Halley's Comet was classified as belonging to Neptune's family.

* The asteroid orbits have similar gaps known as Kirkwood gaps.

However, if one plots to scale a section of the orbital path of Halley's Comet in relation to the orbit of Neptune and Jupiter, it is seen that the comet actually approaches closer to Jupiter than Neptune in spite of the fact that at aphelion the comet lies beyond Neptune's orbit. Nevertheless, all the bulky Jovian planets are to some extent very influential bodies in perturbing comets.

All the short-period comets have direct motions round the

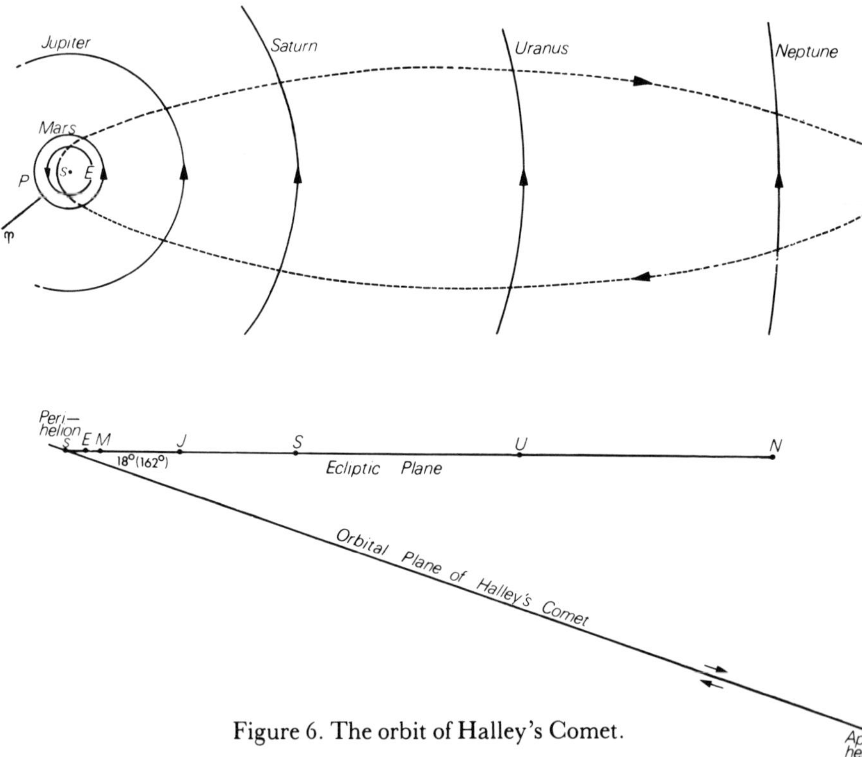

Figure 6. The orbit of Halley's Comet.

Sun with the exception of Halley's Comet and Comet P /Grigg-Mellish, period 164 years. Apart from Comet P /Herschel-Rigollet, period 156 years, these two are among the longer period, short-period comets which have been seen on more than one apparition.

When the comet population as a whole is statistically examined in terms of direct or retrograde motion, half show direct and half show retrograde motion, but there is no immediate and obvious explanation to account for this even distribution.

The orbital elements, when once accurately determined, are literally the fingerprints of a comet, enabling it to be identified from apparition to apparition and, in the case of a new comet discovery, it can be compared with all previous comet apparitions to check whether it is a comet of very long period returning again. Even with short-period comets – whose orbits are continually being altered by perturbing effects of Jupiter and the inner planets – the changed elements can be calculated with great accuracy for any new return to perihelion by introducing certain mathematical criteria.*

During the long history of cometary astronomy, several major comet catalogues have been compiled which enumerate all the comets whose orbital elements have been calculated; and some catalogues also include a list of more doubtful comets whose orbits are not known. The modern catalogues are tabulated in such a way that an orbit computer can conveniently scan through the calculated elements of a comet to compare them directly with the significant orbital elements of all previously known comets.

The comet catalogue published by the British Astronomical Association shows individual entries for over 300 apparitions of 54 short-period comets. Among these were 40 appearances of short-period comets which have made only one appearance and 120 long-period (elliptical) comets of one appearance. In addition are included 300 comets with 'parabolic' orbits and over 70 comets with (problematical) hyperbolic orbits. The parabolic orbits represent comets of very long period, but the comets catalogued with hyperbolic orbits pose a more difficult problem. As it has already been pointed out, a hyperbolic comet implies that it would be entering the sphere of the Sun's influence for the very first time and after reaching perihelion would fly off again into deep space never to return. It might

*For example Tisserand's criterion developed at the end of the nineteenth century.

also suggest an interstellar population of comets travelling in orbits of high velocity which allows them, uniquely, to trip from star to star on single perihelion visitations. But no comets which have been well observed have extremely marked hyperbolic orbits. The highest value is 1·004 for the Comet 1914 III. The Swedish astronomer Strömgren and the Dutch astronomer Mrs van Bilo have shown that comets with hyperbolic orbital characteristics are really periodic comets which have been perturbed into hyperbolic paths by the perturbing effect of a planet in the solar system. These perturbations can have the effect of providing a comet with more orbital energy so that after rounding the Sun at perihelion it accelerates away never to be seen again.

The British Astronomical Association catalogue is the modern extension and revision of the German astronomer Galle's nineteenth century *Cometenbahnen* and Crommelin's two later supplements which extended it. Since this work is only concerned with reliable data, it begins with the apparition of Halley's Comet in 239 BC. However, Chamber's catalogue published in 1910 includes many earlier less authenticated comets beginning with one seen about 1770 BC of which a contemporary remarked: "There was seen a wonderful prodigy in the heavens with regard to the brilliant star Venus . . . Castor avers that this fire star changed colour, size, figure and path: that it was never seen before, and never seen since."

The first astronomer to attempt to form a comet catalogue was Stanislaus Lubienitzki, a Polish astronomer who wrote a two volume work *Theatrum Cometicum* published in Amsterdam in 1668. This is now a comparatively rare work, and the few remaining European copies are rapidly finding their way across the Atlantic into the insatiable shelves of the new American college and university libraries.

Lubienitzki's book was the culmination of what is considered to be the first instance of international cooperation in scientific investigation. After the startling appearance of the comet of 1664–65, he circulated a request to more than 40 astronomers for material in order to carry out his grandiose plans to compile a history of comets.

Unfortunately, however, the scientific content of

Lubienitzki's completed books is unreliable. Many of the 415 comets he compiled were fictitious ones. Lubienitzki led a somewhat obscure life and became subject to persecution by religious enemies. He died in Hamburg after being poisoned. Most of his other works were unpublished, and nearly all his manuscripts are lost or destroyed.

The great observational astronomer Hevelius wrote a history of comets which is preserved in his 12 books *Cométographie*. But the first truly critical work was written by Nicolas Struyck who published *Algemeene Geographie* in Amsterdam in 1740. His later book *Vervolg van de Beschryving der Staatsterren*, 1753, was the genesis of the French astronomer Pingré's great *Cométographie*.* This work, published in about 1784, is still the most extant work on the subject of cometary history and is never likely to be surpassed.

Pingré was born in Paris in 1711 and at the age of 16 entered a scholastic order. By the age of 24 he held the chair of theology, but a purge by the Jesuits deposed him. It was not until he was 38 that he was offered an astronomical position. This he gratefully accepted and he set about the task of re-educating and equipping himself for his new interests. Like Halley he made a number of astronomical sea voyages. Later in life he was able to return to his old university where he finished his monumental work. The book is a veritable fountainhead of all aspects of cometary astronomy up to his time. Perhaps its most significant contribution was the examination of the orbit of every comet which could be traced. In recent years the Comet Commission of the International Astronomical Union has discussed ideas for producing a modern extension of the *Cométographie*, but the problems involved in the compilation, production and printing might well prove to be an economically insurmountable obstacle unless a generous institutional benefactor can be found to underwrite the whole venture.

The Chinese were greatly interested in comets, and reference has already been made to their scientific methods in observing them in contrast to the semi-hysterical approaches of contemporary Europeans. The first of the long series of Chinese observations came to light through the researches of

* Full title *Cométographie ou Traité Historique et Théorique des Comètes*.

Jesuit missionary priests in Peking early in the eighteenth century. In 1846, the great French astronomer-cum-Oriental historian E. Biot published a translation of Chinese comet catalogues. In 1871, the Englishman J. Williams published an elaborate scholastic work which became the extant source of Chinese comets, and it remains so to this day.

1. The 1066 return of Halley's Comet depicted on the Bayeux Tapestry.

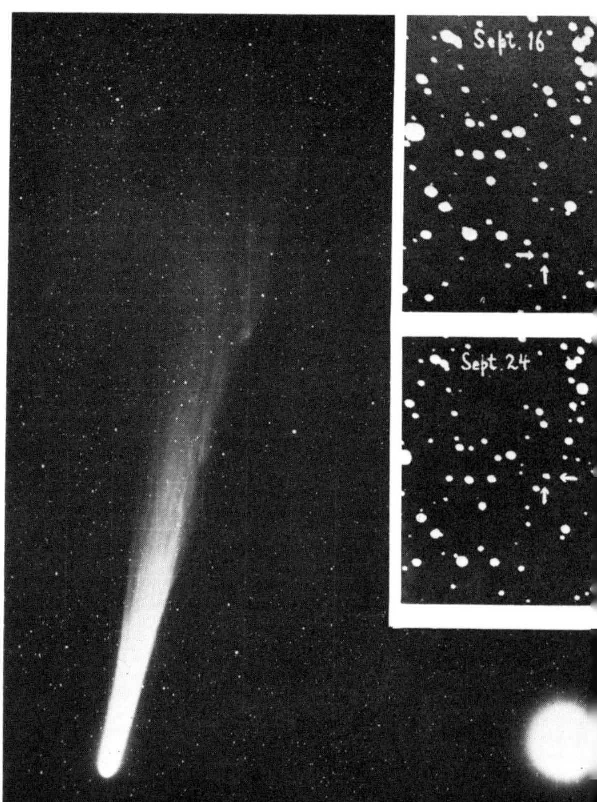

2. Halley's Comet in 1910. The bright object to the right is the overexposed image of the planet Venus. *(Inset)* Halley's Comet 16th and 24th September 1909 on its approach to the Sun after 74 years.

3. A contemporary impression of Donati's Comet over Paris in 1858 (with bright star Arcturus to right of head). The magnificent 'dust' tail appears like a scimitar, while the two gas tails are straight, narrow, pencil-beam features.

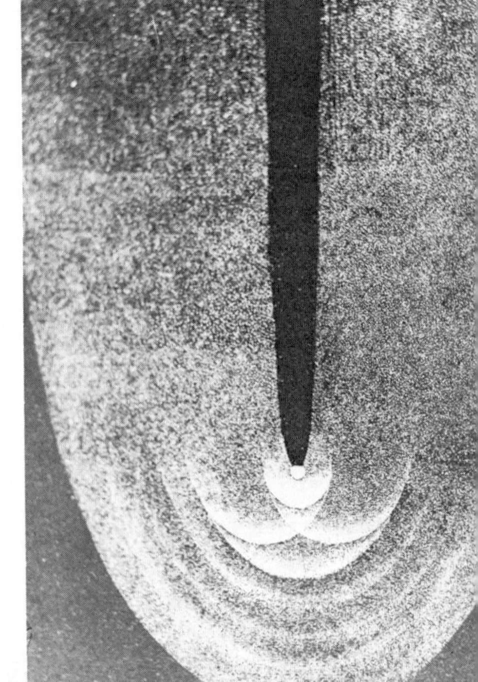

4. Coggia's Comet in July 1874, showing expanding envelopes moving out from the nucleus.

5. The Great Comet of 1744 depicted by a contemporary artist showing its six tails projecting above the horizon.

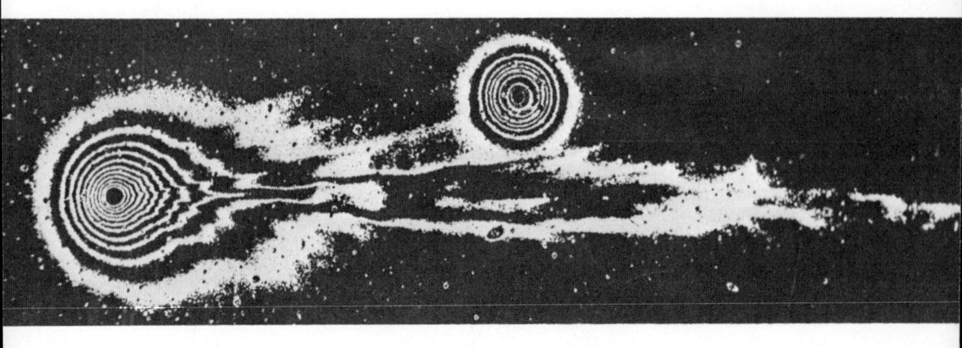

6. Comet Whipple-Fedka-Tevzadze (1943 I) showing the extent of the ionized carbon monoxide (CO^+) extending into the tail. The circular object is a bright star.

7. Comet Arend-Roland (1957 III) on 24th April 1957. Note the remarkable sunward spike.

8. Meteor trail and Comet Brooks (1893 IV) on 13th November 1893.

9. Comet Mrkos (1957 V) on 21st August 1957 shortly after discovery with the naked eye.

10. The Great Comet of 1861 on 30th June as the Earth passed through the comet's tail.

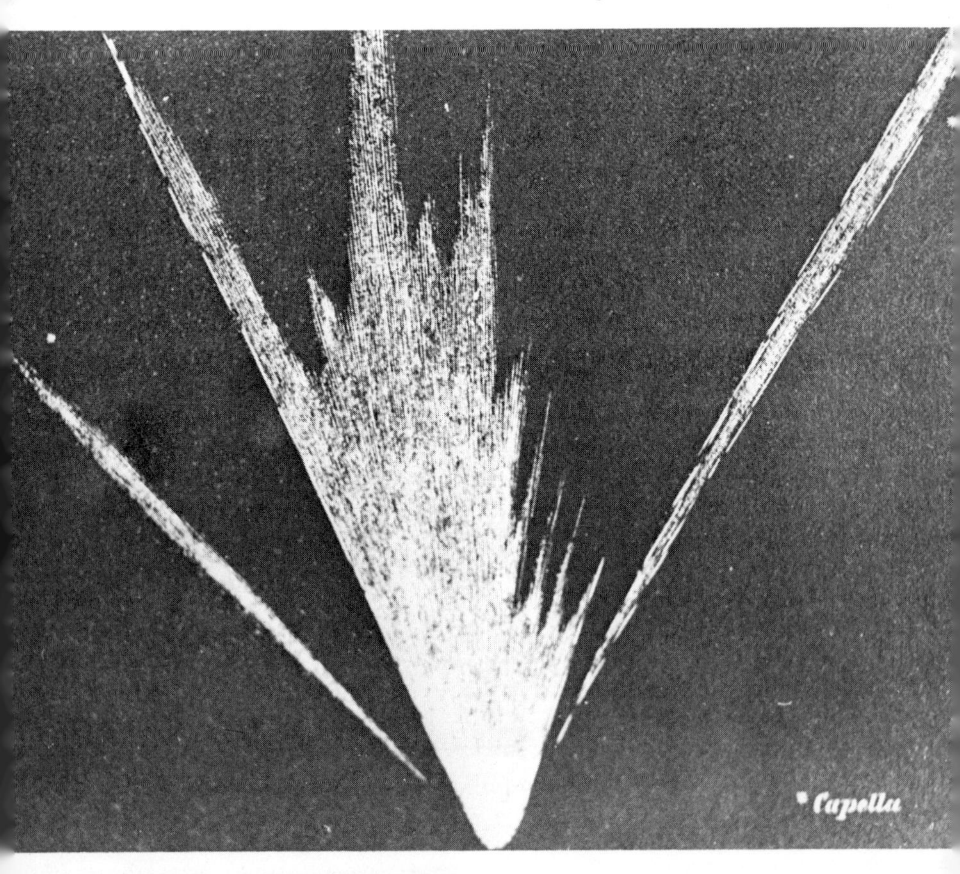

CHAPTER IV

Nature of Comets

It may be remembered that earlier – in the introduction – a comet was physically defined as a cloud of meteors, surrounded (at times) by a gaseous envelope . . . This simple definition, however, is only a generalized description, for at the present time there is no universal agreement among astronomers about the detailed, structural picture concerning the nature of comets.

Although the fundamental mathematical problems of cometary orbits were solved in the time of Newton and Halley, no immediate solution was forthcoming in the period which followed about the nature of comets, nor could anyone offer an acceptable argument to explain their presence and origins.

Structurally a bright comet may consist of four distinct parts: the nucleus, coma and tail plus a fourth highly tenuous component discovered in 1970, but which is not directly observable from the surface of the Earth. The object of real mystery is the nucleus – which from an observational point of view is the least spectacular – since it is still permissible to ask whether such a discrete object actually exists in the head of a comet.

Some comets *do* show a distinct star-like object usually near the centre (occasionally displaced) which despite the highest practical magnification applied to a large telescope will not resolve and expand itself into anything larger than a point source. Other comets show no such discrete star-like body enveloped in the coma and present the appearance of a concentrated cloud of nebulous material which gradually shades off at the edges so that it is difficult to define its actual boundaries.

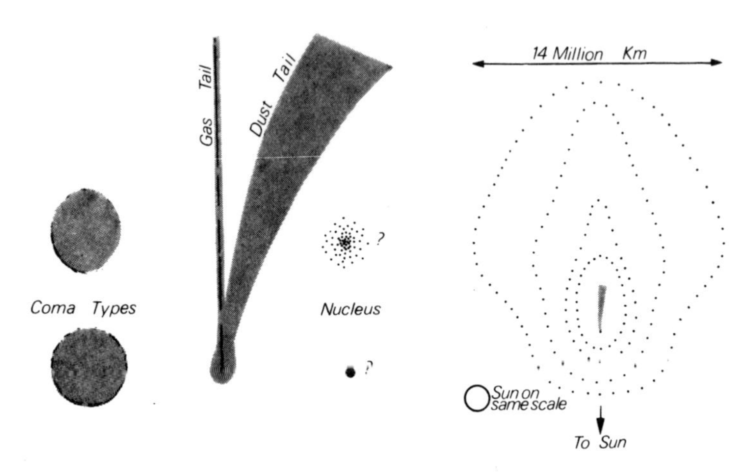

Figure 7. The structural parts of a comet: figure (right) shows the limits of the hydrogen gas envelope of Comet Bennet 1970 II as scanned from OGO-5 in Lyman-alpha light (1216 angstroms). The different contour boundaries represent different densities of tenuous hydrogen gas surrounding the comet (shown in the centre).

Some observers in the past claimed to have resolved discrete nuclei into circular planet-like discs, but in modern times there is not a shred of reliable observational evidence to confirm these earlier claims. Photography offers little assistance in settling the question – apart from setting an *upper limit* to the size of nuclei if they exist. Work at the US Naval Observatory by Elizabeth Roemer using a 40-inch astrometric reflecting telescope of very high definition has shown that comet nuclei cannot be *larger* than about 3 to 4 kilometres for short-period comets and 7 to 8 kilometres for long-period comets.

One of the closest approaches by a comet in modern times occurred in 1927 when the P/Pons-Winnecke Comet came within 5.6 million kilometres (3.5 million miles) of the Earth. The French astronomer Baldet examined it under very high magnification with the 30-inch refracting telescope at the Meudon Observatory, and he came to the conclusion that the nuclear-like bright point at the centre could not be more than one kilometre in diameter – if indeed it could be considered a single discrete body at all.

One of the most powerful tools of modern science is the spectroscope, and when first developed in the middle of the nineteenth century, it was applied to comets in the expectation it would provide a Rosetta stone in deciphering their peculiar nature and chemistry.

It was in August 1864 that the Italian astronomer Donati – discoverer of the brilliant comet of 1858 – made the first visual observation of a comet's spectrum. He had just adapted a small spectroscope to begin a programme of star spectra observations, and conveniently Tempel's Comet 1864 II suddenly appeared which, although faint, was ideally positioned in the sky. In recording his observation Donati wrote: ". . . the dark parts are larger than the luminous part(s), and one might say that these spectra are composed of three bright rays . . ." What he had in fact observed were the bright Swan bands of carbon which he also noted resembled the spectra he had obtained from metals in the laboratory.

By a remarkable coincidence the next cometary spectrum to be observed was in another of Tempel's comets, this time 1866 I.* On this occasion it was also seen by the Englishman Sir William Huggins, one of the most inspired spectroscopists of the nineteenth century, and it was observed spectroscopically by the eminent Italian astronomer Secchi. Both observed the three bright bands seen by Donati, and in addition they recorded the Fraunhofer (continuous) spectrum due to reflected sunlight.

In the period 1865–71, Huggins examined the spectra of a further six comets, none of which were outstandingly bright objects, and consequently their spectra could not be resolved in fine detail. However, all these observations helped towards building up theories about the chemistry of cometary heads. . . .

When the light of the nuclear region is examined, the spectrum reveals a very distinctive Fraunhofer series of lines which originate as *reflected* light from the Sun. By contrast the spec-

* The comet is connected with the annual meteor shower known as the Leonids (see page 203).

trum of the surrounding coma region shows the strong emission lines and bands of luminous gases. In some comets the Fraunhofer (solar) spectra dominate the entire cometary spectra while in others the emission lines dominate – indicating that comets may differ radically in their physical make-up.

The emission lines and bands – some of them highly complex structures – are due to radiation by gas molecules and probably originate from parent molecules* such as cyanogen (C_2N_2), carbon (C_2), methane (CH_4), ammonia (NH_3), carbon dioxide (CO_2), carbon monoxide (CO), nitrogen (N_2), sodium (Na) plus compounds of oxygen and hydrogen. In addition, calcium, iron and nickel have been detected in emissions from the nuclear region in two of the Sungrazing comets when they were near the point of skimming the outer, tenous atmospheric regions of the Sun.

Cometary emission spectra are far from being fully understood, for there remain many emission lines that cannot be identified.† One difficulty is that the continuous Fraunhofer spectrum of the cometary nuclear region is superimposed on top of the gaseous emission spectra originating from the coma, and emissions are simulated which in reality do not exist at all!

Many astronomers prior to the development of the spectroscope considered that comets were self-luminous bodies similar to the Sun, yet paradoxically they also thought them habitable bodies like the planets.

In modern times three mechanisms have been short-listed as possible contributors to comet luminescence – apart from that portion of a comet's light which is reflected sunlight.

a) Electric collision
b) Photo-dissociation by solar radiation
c) Fluorescence (excited by solar radiation)

* The dissociation of parent molecules in the comet's head into daughter molecules is a direct consequence of solar radiation, but the parent molecules themselves are difficult to identify since they emit radiation in the part of the spectrum (the red and ultraviolet regions) not available for study except from outside the Earth's atmosphere.

† The daughter molecules (which are often ionized) are chemically unstable and readily combine, and owing to the very low gas densities which are encountered in the cometary coma, they exist in forms unfamiliar on earth.

Nowadays there is no doubt that fluorescence is the most likely principal contributor, since collisions between molecules and electrons must be practically non-existent in the coma owing to the very rarefied gas densities (remember a "bagful of nothing" definition) and the main role of photodissociation (in comets) lies in the transformation of parent molecules into daughter molecules. But the source of the gas which gives rise to the emission spectra is yet another controversial issue. Indeed, because of conflicting ideas encountered when one is discussing the physico-chemical nature of comets, it is necessary to refer to particular theories – or models as some prefer to term them.

The 'Dirty Snowball' Model

The genesis of 'the dirty snowball comet model' has its beginnings in independent suggestions put forward by the German Hirn and the Englishman Ranyard in the nineteenth century. After World War II it was revised and elaborated on by the American Fred Whipple and the Soviet astronomers Vsekhsvyatsky and Levin.

Whipple in particular has put forward a highly descriptive theory to show how this model accounts for the observed physical behaviour of a comet. The whole basis of his idea is the *a priori* acceptance of a solid nucleus in the centre of a comet that is visualized as a very porous mass of solidified gases (or ices) made up of water (H_2O), ammonia (NH_3), methane (CH_4), possibly carbon dioxide (CO_2) and dicyanogen (C_2H_2) which also includes occasional solid particles. Whipple has picturesquely likened his comet nucleus as being analogous to a mixture with the consistency of a yeasty raisin bread.

Now such a spongy mixture of material would be a poor conductor of heat and although the outer surface would be heated by solar radiation – particularly when the comet was at perihelion – the inner material would be warmed very slowly. From the well-known properties of several solidified gases, one could assume that methane would be the only one to vaporize at a solar distance of several astronomical units as the comet approaches the Sun along its highly elongated orbit. When it nears the orbit of Mars, carbon dioxide and ammonia would evaporate, and nearer the Sun, dicyanogen and water would become gaseous. Also through the absorption of sunlight the action of photo dissociation will reduce the gases into less complex molecules. This speculative reasoning is in broad agreement with the changes observed in the spectrum of the coma.

A slight variation of this comet model is one proposed by the Esthonian astronomer Öpik who in fashionable, colloquial language has nicknamed it "the layer cake model". This elegant description is not without good reason, for it advances the idea of a cometary nucleus built up in alternating layers of ices and meteoric particles analogous to some terrestrial sedimentary rock formations.

The ices would evaporate in natural succession – according to the increasing evaporating temperature – as the comet makes its perihelion approach. The most volatile substances (such as hydrogen, nitrogen and methane) would boil in the deep cold layer and at smaller internal depth they would vaporize and disappear. No melting would take place only evaporation in the solid state thus ensuring porosity in the matrix and a free escape of gases from different layers. The main difference between his model and Whipple's is that Öpik reckons that at the centre of the nucleus there may be a chunk of solid hydrogen constituting a core which initially formed at the lowest temperature of interstellar space and is preserved inside the comet at a temperature of 5 to 10°K.

Whipple's and Öpik's fundamental idea of a discrete icy-conglomerate comet model has received wide acceptance among many contemporary astronomers. However, this idea poses other unresolved questions such as:

a) Is a comet nucleus made of one solid chunk of ice, a few blocks, many small solids or puffy ice?
b) Are the ices formed round minerals and meteoric dirt, or are the solids randomly distributed in the ices?
c) Is the nucleus a ball, like the planets, irregular or rough?
d) Does the nucleus rotate?
e) What are the differences in temperature of the dark and light sides?
f) Does the nucleus possess any magnetic properties?

The 'Flying Sandbank' Model

'The flying sandbank' is another picturesque label given to one of the alternative models which is a contemporary updating of the nineteenth century concept of a comet. In modern times the Cambridge astronomer Lyttleton has championed it and he has been able to explain fairly satisfactorily many of the former objections to this model.

In the Lyttleton theory the nucleus and the coma of a comet are imagined as a continuous structure in the shape of a gigantic cloud of widely-scattered dust particles whose mean distance (between each particle) is extremely large, but which in the centre becomes more concentrated; thus giving rise to the illusionary appearance of a solid monolithic body at the comet's centre.

The idea supposes that each comet particle describes an independent orbit round the Sun, but they traverse a path round the Sun in such a way that their grouping orbits are the same as the comet's orbit.

To the visual observer the Lyttleton picture – if not wholly acceptable – is a reasonably convincing one. The visual observer is used to seeing a comet as a faint, diffuse cloud-like smudge with very little suggestion of the dense icy-conglomerate ideas often misleadingly portrayed in over-exposed photographs of comets. However, some bright comets *do* appear to be highly concentrated at their centres, more so than one might expect them to be from the illusionary concentration of particles. In addition many of the brighter comets show jets, hoods or fountains of luminous material which appear to originate from the centre of the comet, and it is easier to reconcile these phenomena with the idea of them emanating from a solid mass forming the nucleus. Nevertheless, many of the fainter comets appear to have no nucleus at all and even when observed at close quarters, they remain diffuse cloud-like objects.

The truth may well be that some comets do have discrete monolithic nuclei, and others have none. One of the attractions of the Lyttleton model is that it is able to offer an all-embracing theory of comets which includes the *modus operandi* of their formation. On the other hand the Whipple model does not set out to explain the origin of comets and needs to be stitched to other ideas which do.

Nevertheless, at the present time, it must be emphasized that *all* ideas regarding the nature and origin of comets are highly speculative ones.*

Some of the most puzzling features of comets – particularly the brilliant comets – are the previously mentioned jets or hoods of luminous material which appear to be ejected from the nuclear region. They often develop into expanding envelopes which change their appearances from hour to hour. During the 1835 apparition of Halley's Comet, John Herschel wrote of his observations on 28th October:

> . . . Within the well defined head, and somewhat eccentrically placed, was seen an individual nucleus, or rather an object which I know no better way to describe than calling it a miniature comet having a nucleus, coma and tail of its own, perfectly distinct and exceeding in intensity of the light the nebulous cloud or envelope...

During this apparition the comet proceeded alternately to lose or gain a tail and occasionally contract until it resembled a star; then it would expand outwards to form a nebulous globe which then gradually faded and was lost again. The German astronomer Bessel hinted that it must be due to an emission of powerful electrical forces. On 22nd October 1835 he noted that the nucleus suddenly appeared to have acquired a striking new brilliance from which there emerged, on the side closest the Sun, a cone of light which after extending to a short distance from the head, was observed to curl backwards as if impelled by a force of great intensity directed from the

* All science is of course speculative: According to one of the basic tenets of the philosophy of science as formulated by the celebrated Karl Popper, we can never be sure that any particular theory is correct. A theory is successful as long as it stands up to the tests. What we can say is that one theory has more 'verisimilitude' than another, but we cannot say any theory is an established absolute truth describing a phenomenon.

We dress up a theory in mathematical and verbal language which at best can only describe close approximations to the truth. Some famous examples are Newton's "laws" and Einstein's theory of relativity which are in reality only working hypotheses to assist us in making sense (or use) of what we observe. This in no way detracts from the traditional ideals of the scientific pursuit for knowledge – instead it highlights the ideas behind the scientific method which is to eliminate the unlikely ideas and narrow the field of speculation rather than chase the unknowable absolutes.

Sun. From night to night thereafter it was seen to vary constantly both in brightness and in size, and in addition the axis of the coma was also variable.

Many of the brilliant comets observed in history have shown these characteristic luminous jets and fountains. Comet Bennet 1969*i* was seen at one stage to have some half-dozen separate jet-ray streamers emerging from the nuclear area.

The problem of recording the fine details of these jets is not as simple as it may appear. First thought is that photography might be able to offer the most definitive way, since visual drawings made by observers at the eyepiece of a telescope are highly subjective. Unfortunately, many comet photographs are either over-exposed in order to record maximum length of tail or under-exposed to show only the brighter parts of the inner coma to ensure accurate plate measurement. Some photographs have recorded these jets, but not in the same fine detail that can be seen visually even with modest-sized telescopes.

Comet Tails

The tail of a spectacular comet has always attracted public attention in the night sky, and in some instances when a comet is suitably placed, the tail appears to stretch right across the heavens (Plate 13). The tail of a great comet is probably the most voluminous thing in the solar system, but not all comets have tails. We can go even further and say that tails, in particular the spectacular ones, are celestial rarities. Sometimes when one examines a comet telescopically, it is difficult to say where the coma ends and the tail begins. Many of the short-periodic comets never show a tail as anything larger than a faint elongation of the coma – perhaps barely half a degree in length and only then visible on photographs.

From descriptions by astronomers in the ancient Western World it is often difficult to decide whether some of the so-called spectacular comets of the past were, in fact, comets at all. There can be no doubt that many records of past cometary apparitions which described the phenomenon as having the colour of blood and which changed rapidly in form, were more likely auroral displays. Fortunately for cometary astronomy, the ancient Chinese astronomers took very careful note of comets so that their physical observations were more definitive. They nicknamed them 'broom stars' to distinguish them from 'guest stars' or exploding stars.* It was the ancient Chinese observers who first noted in AD 837, that cometary tails always appear to be directed away from the Sun, and it was not until Peter Apian remarked upon it in 1531, that Europeans became aware of it as a general rule. We can read that Tycho Brahe did not believe that this had been sufficiently well demonstrated and refused to accept it.

Comets are always brightest when near the Sun. When a large comet approaches the Sun, it develops its tail which gradually grows longer and becomes very much brighter. If observed in the western sky at sunset or in the eastern sky before dawn, the tail of a bright comet will appear to point upwards. When it swings round the Sun at time of perihelion, the tail also swings round in such a manner to precede the comet on its journey back into space. Generally speaking, comets, when receding, have longer tails at the same distance from the Sun than when approaching at the same distance before perihelion owing to a delayed action in the maximum physical development of the tail.

It is evident that the mechanism which is the cause of the tail lies somewhere in the coma and/or nucleus of the comet and since they always appear to point away from the Sun, there must be present in space some kind of repulsive force (or forces) directed away from the Sun. Because of this, any particle of matter forming part of the tail is subject to at least two distinct forces: one caused by the solar gravitational attraction and one by the solar repulsive force which repels it. Although it is assumed that the tail always lies in the opposite direction to the Sun, this is not strictly correct with every bright comet. Sometimes there appears to be a definite lag in the tail behind the Sun/comet line (the radius vector), in some instances multiple tails can be observed spread at wide angles to each other. But the angle at which a comet tail is observed depends on the relative position of the Sun, comet and Earth; the effects of perspective can be highly misleading to the observer, especially so if the comet is passing close by the Earth.

* Novae and supernovae.

The first theoretical studies of the tail formation processes were made by Isaac Newton. Among earlier speculations about comet tails was one put forward by the philosopher Panaetius who held the view that they did not exist in reality but were a kind of mirage caused by the rays of the Sun, and this belief continued until the seventeenth century. Both Hooke and Newton assumed a repulsive force at work, but it was left to Newton to vaguely account for it – with notably less success than his theories about gravitation.

It was not until Bessel (1784–1846) began to think about the problem that more scientific theories were formulated, and it was Bessel and his fellow countryman Olbers* (1758–1840) who first began to think in terms of electrical forces at work. The Russian astronomer Bredichin (1831–1904) advanced the idea of three kinds of tail to explain the various tail shapes assumed by comets based on the idea of different gases, which at the time served reasonably well to explain matters. Earlier, at the beginning of the seventeenth century, Kepler and Euler had suggested that the pressure of sunlight might be a contributory factor in the production of the tail. Three hundred years later (1900–01) the physicists Lebedew, Nichols and Hull demonstrated the existence of such a force in the laboratory. Thereafter solar radiation pressure was acknowledged to be the principal agent in tail formation.

With the development of spectroscopy it became possible to study the chemistry of comet tails, and it was found they consist of both gas and dusty material.

For a time the solar radiation pressure theory appeared to be wholly satisfactory in accounting for the origin of *all* tails. However, in 1908, Morehouse's Comet appeared which behaved in a highly unusual fashion. This comet displayed great activity, and knots of material were ejected from the head and travelled down the tail at speeds far in excess of what was considered reasonable on the assumption of the solar radiation pressure theory. Sir Arthur Eddington,† then employed as Chief Assistant at the Royal Observatory, Greenwich, studied Morehouse's Comet over the whole period of its

* A famous German amateur astronomer who also refined and simplified the method of computing parabolic orbits.

† Later famous for his cosmological ideas.

visibility. Eddington's earliest astronomical success was in accurately measuring the velocity of the ejected material. This he found was travelling at the speed of 112,000 kilometres (70,000 miles) per hour, representing a solar repulsion force of 800 times larger than the pull of gravity! Brooks' Comet 1893 IV had shown similar turbulent motions in the tail, but Morehouse's Comet was the first one to be studied intensively by photography over an extended period.

It is possible to explain forces of 20 to 30 times solar gravitational attraction by radiation pressure alone, but radiation pressure will certainly not account for the great repulsions in Morehouse's Comet. Nowadays, although radiation pressure is still considered a principal contributory factor in the production of cometary tails, an additional force has been proposed to account for the high velocities in gas tails, which is in harmony with contemporary ideas in physics. Modern astrophysical research has revealed the existence of solar corpuscular radiation which consists of streams of protons and free-electrons ejected into space. These particles, known collectively as the solar wind, travel at speeds ranging from 1,000 to 2,000 kilometres per second, and there is also a slower faction which travels at 500 kilometres per second.

According to a theory first put forward in 1951 by the German astronomer Biermann, these particle streams are very likely the principal cause of cometary tails which result due to the interaction between the cometary head and the solar particle streams meeting it. A similar idea was put forward by the American Schaeberle in 1893, but physical science could not then offer a convincing explanation.

Tails may be distinguished into three types retaining the older Bredichin nomenclature, but for different reasons. There are tails which are caused by interaction with the solar wind particles that produce tails consisting of ionized gases (or plasmas),* and tails which consist of a variable mixture of dust and quasi-dust particles resulting from radiation pressure.

Nowadays, Bredichin's original three-type classification is represented by Type I gas tails, due to ionized gases such as carbon monoxide (CO^+), and Types II and III consisting of a

* Plasma is sometimes described as "the fourth state of matter".

mixture of dust and non-ionized gases for the tails which are in accord with the radiation pressure theory. The pure dust tail is strongly curved like a scimitar (and tends to lag behind the radius vector), and the ionized gas tail is generally narrower and very straight.

If one looks at a photograph of a comet, the different tails are easily recognizable. In many instances some comets possess both gas and dust tails. The famous Donati Comet (Plate 3) showed a magnificent tail plume of what we now know to be 'dusty' material and also two narrower, almost straight tails of ionized gas. Modern photographic studies show many similar examples. The short-periodic comets generally have narrow (photographic) tails which chiefly consist of ionized carbon monoxide – while the long-periodic comets have spectacular tails showing a continuous Fraunhofer spectrum, thus indicating that the reflected light is due to the presence of solid or quasi-solid dust particles.

As mentioned earlier, old records reported comets the colour of blood, and reference has already been made that these 'comets' were more likely auroral events in the upper atmosphere of the Earth. However, comet tails may at times be genuinely coloured, although it is a very marginal effect. Occasionally a bright comet near the Sun appears to glow with a definite shade of yellow, and this can be directly attributed to the effect of emission of sodium light in the coma which is only triggered off at small perihelion distances. The great southern Comet (1947 XII), discovered by British sailors in the South Atlantic in 1947, and the 1957 III Arend-Roland Comet were both quite yellow. When other colours are represented, such as reds, greens and blues, they can usually (but not always) be attributed to meteorological effects in the Earth's atmosphere at the time of observation.

Generally, comet tails are described as being highly transparent, and stars can be observed through them without any diminution of brilliance. In spite of their volume the mass of a tail must be infinitely small. When the tail of Halley's Comet reached its greatest length in 1910, Schwarzschild estimated that if it were composed of only fine dust particles, it might possibly weigh 1,000,000 tons, but if it were gas, it would be considerably less – perhaps not more than 1,000 tons. We know now that both kinds of material are present; if we use the larger figure, which still sounds like a great deal of material, and take into account its dimensions, this would imply that it contains one cubic inch of air at atmospheric pressure disseminated throughout a volume of 200 cubic miles! No modern vacuum-pumping technique can hope to simulate this tenuous density on Earth.

It can be seen that if the Earth were to pass through the tail of a comet, we should expect to notice little effect. During the Earth's long history this must have happened quite frequently, and it occurred as recently as in 1910 with Halley's Comet. Many scientific theories of the past have been engendered to explain some peculiar event which took place during the Earth's history by its passage through the tail of a comet, but there is no evidence to suppose that such an encounter could have the slightest physical influence on the Earth, although the encounter could result in a considerable influence on the comet itself.

Comets, on occasions, have been reported with multiple tails. The most famous example is De Chéseaux's Comet* of 1744, which was depicted on contemporary drawings with six distinct tails (Plate 5). But modern photographic methods have recorded comets with fainter tails sometimes double this number. The great comet of 1861 (which is another example of the Earth's encounter with a tail without apparent effects), showed a variety of ray-like features at the time of the Earth's passage through it. The large 'dust' tails which emerge in sweeping fountain-like parabolas forward from the cometary head – which are pushed backwards as the solar repulsive force overcomes the propulsive forces – are probably cylindrical in structure with comparatively hollow centres, this being brought about by the shielding effect of the cometary centre or nucleus. There is some evidence to show that both gas and dust tails may rotate, but there is great difficulty in recording direct evidence with present-day photographic techniques. Methods using cinematographic techniques are at present being tried out to test the theory of tail rotation.

Some of the more interesting tails are those reported by the Chinese as belonging to 'bearded' comets which gave the

* More correctly Klinkenberg's Comet (see page 82).

appearance of a tail *ahead* of the comet apparently reaching out towards the Sun. In modern times such a tail was observed in the Arend-Roland Comet (1957 III), which on 26th April 1957 showed a bright spike-like tail apparently directed exactly *towards* the Sun (Plate 7). This aroused considerable interest and excitement among observers, but it was soon realized that the appearance was simply one of perspective due to the Earth passing through the comet's orbital plane (Figure 8).

Apart from the ancient Chinese reports of the same phenomenon, similar observations were also reported in Comet

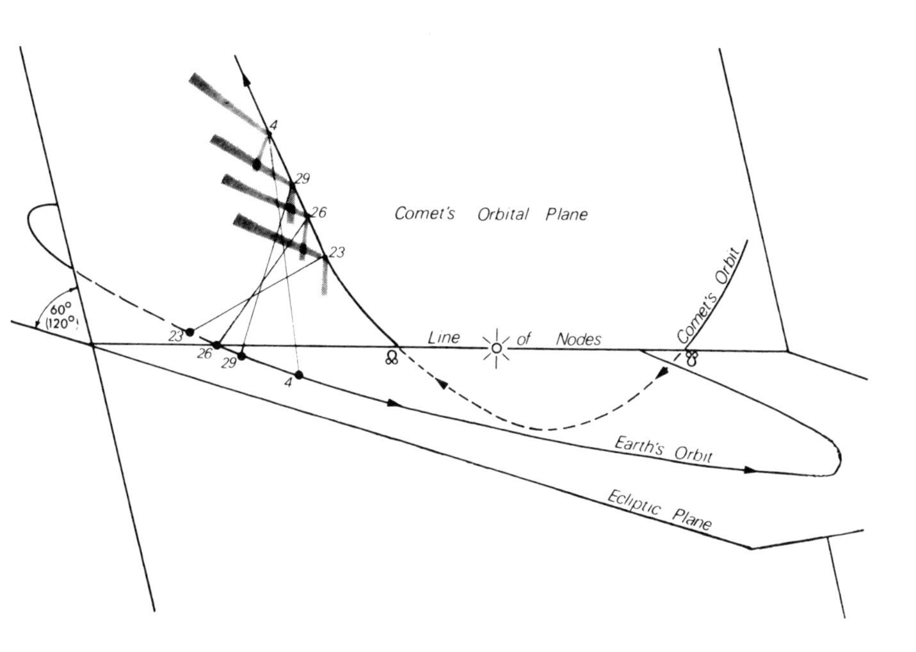

Figure 8. The orbit of Comet Arend-Roland 1957 III showing the position of the Earth and the comet on the 23rd, 26th, 29th April and 4th May 1957. The Earth passed through the comet's orbital plane about 26th April, and on a few days before and after, the so-called (anomalous) sunward tail spike was seen. This spike was due to fine dust lying in the orbital plane seen edge-on, and its apparent sunward extension was simply a geometrical illusion.

1862 III, the one associated with the annual Perseid meteor shower (see page 208). There can be no doubt that these 'sunward spikes' are simply the effect of the illumination of fine 'dusty' material lying in the plane of the comet's orbit. It may be supposed that all cometary orbits have associated 'dusty' material, but the comet is rarely so uniquely placed that observers can see it.

Theories about the nature of comets must also account for all manner of associated problems. One of these is concerned with the life spans of comets and in particular how some of the short-period comets are able to go on producing gaseous envelopes which are then swept away by the ever present solar wind and lost each time they come to perihelion.

In the Whipple 'dirty snowball' concept of a comet the idea is put forward that comets have very limited life spans. Predictions about the 'death dates' of well-known periodic comets have been made on the assumption that after each perihelion approach the comet grows dimmer. The Soviet astronomer Vsekhsvyatsky (who also has some unusual ideas of comet origins, see page 75) has attempted to show that many of the short-periodic comets are *rapidly* growing fainter. This evidence, however, is based on some very doubtful observational material. In particular one of the comets cited, P/Tempel 2, which Vsekhsvyatsky showed to be diminishing rapidly, was observed at the 1967 perihelion approach, after 14 revolutions round the Sun, to be as bright (if not slightly brighter!) as when it was first discovered in 1870. Similar claims for the rapid fading of Encke's Comet (page 95) are also made, but if Encke's Comet is fading, it cannot be more than about one magnitude per century.

The Soviet astronomer Levin – a proponent of the 'snowball model' – accounts for the behaviour of the gaseous envelope by applying to the problem well-known ideas about the physical theory of gases. As applied to comets it is visualized that the gas molecules are somehow attached to the solid particles in the nucleus or nuclear region. In chemical terms this is called the *adsorbed* state. It is based on a well-known concept that an absorbing material covers itself with a densely packed layer of gas or liquid, and the skin becomes very strongly attached to the surface. The important application to comets is that the power to absorb the gas decreases with an increase in temperature and increases if the temperature is lowered. The decrease process is known as desorption.

The ability of adsorption is dependent on the surface of a substance. A porous material such as charcoal is able to adsorb a surprisingly high quantity of gas. Making use of this property, Levin has demonstrated that it is possible to explain the apparently vast reservoirs of gas locked up in a comet. Whether solid cometary particles have the ability to replenish their store of gas at distances far removed from the Sun is not known. However, it is known that out in the colder regions of space, away from the immediate solar influence, there are vast amounts of tenuous, dusty and gaseous material that might well be gravitationally attracted or collected when a body such as a comet passes through it.

At the present time a great deal of rethinking is taking place among cometary astronomers about the physical and chemical processes that take place within the coma. It has long been recognized that knowledge of the physical processes taking place will never be known with certainty from earthbound observations. It was long realized that the observation of a comet from outside the Earth's atmosphere may radically alter our ideas about them.

The first opportunity for such an observation occurred in January 1970 when the Orbiting Astronomical Observatory 2 (OAO2), launched on 7th December 1968, first observed Comet Tago-Sato-Kosaka using an ultraviolet telescope. It made the interesting discovery that the comet was surrounded and enveloped in a vast hydrogen cloud about a million miles in diameter (or a little more than the diameter of the Sun).

In April 1970, observations from the Orbiting Geophysical Observatory 5 (OGO5) made two ultraviolet scans of Comet Bennet – one of the most brilliant comets of the century – and found it enveloped in a hydrogen cloud over eight million miles in diameter.

These cloud features can only be seen in ultraviolet light which is normally absorbed by the Earth's atmosphere. The observations indicate that we still have much to learn of the physical mechanisms at work in comets, and until a space probe is dispatched to explore the inner coma of a comet, the problems will doubtlessly remain unsolved. At the time of writing, in the early 1970s, many such probes are planned for the future (see page 148).

CHAPTER V

Origin of Comets

The presence of comets, and their significance, has occupied the attention of astronomers since earliest times. Although the Babylonians are reputed to have recognized their true astronomical character, we only have, as confirmation, very slender apocryphal hearsay evidence. Later Aristotle is quite emphatic about his own ideas, and thought comets to be simply meteorological events rendered visible in the atmosphere. Historically Seneca, in his *Questiones Naturales*, provides the most extant viable scientific 'guess' when he wrote that he considered them to be somehow akin to the wandering planets but with different paths. The birth of modern scientific speculation about comet origins began in the early nineteenth century when Laplace (1749–1827) first suggested that comets had their origin in an interstellar cloud captured by the Sun. Shortly after, Lagrange (1736–1813) expressed an opposite view in declaring that all the evidence indicated they originated as material which had been violently expelled from the major planets. Today both these views are still championed in cometary cosmology in addition to some more modern variants.

During the nineteenth century the interstellar ideas of Laplace were much favoured, particularly from a descriptive point of view. It was not until the idea was examined more critically that great doubts began to be cast. It was argued that if a comet were captured from space, its subsequent orbit should retain tell-tale indications of this.

Laplace had supposed that comets were composed of

nebulous matter originally surrounding the solar system at an assumed distance of 100,000 astronomical units. Some of the comets arriving to within observable distance would have hyperbolic velocities. The idea received a set-back when it became known from which direction the Sun appeared to be moving in space. It follows as a consequence of Laplace's theory that we ought to meet more comets than those which overtake us, and many of those we meet should be markedly hyperbolic. In practice this is not the case with any comets *well observed* within the modern era. Although statistical evidence indicates that the axes of cometary orbits – excluding those of short period – are randomly scattered and show no marked distribution excess in any particular direction, it is only true as a generalization. Some of the comets of very long period show a slight preference for the plane of the galaxy. More significantly the Sungrazer group of comets shows evidence for a direct link between their direction of perihelion and the direction of the Sun's motion in space. There is good reason to suppose this group of comets originated from a single parent body which subsequently split up (see page 88). Other families of comets with common orbital characteristics may also have formed from single progeny 'mother' comets.

Until Lyttleton published his ideas in the early 1950s, the interstellar origin of comets had been practically superseded by ideas invoking mechanisms for comet origin within the solar system. Lyttleton set out to explain both the nature and origins of comets, since their nature is a consequence of their origins.

Instead of assuming as Laplace had done that comets were condensations already in being, lying beyond the periphery of the solar system, Lyttleton considered that the process of condensation and formation of a comet took place near the Sun. He also assumed the ability of the Sun to attract particles of dust and gas probably extending to a distance of 1,000 AU, although others subsequent to Lyttleton consider that 1,000,000 AU may set the true limit. As a consequence of the *theory of accretion* – formulated by Herman Bondi and Fred Hoyle in the early 1940s – Lyttleton extends and adapts it to account for comet formation. In simple descriptive terms it considers what processes may take place when the Sun passes through a homogeneous cosmic cloud of dust and gas particles. Lyttleton considers that the Sun produces condensations within it and subsequently impresses on these condensations the orbital forms of comets.

The result of the action of the Sun would be to form an accretion axis in the direction opposite the Sun's way in space (the anti-apex*) where the transverse velocities of particles might be de-energized by collisions. Then over a period of time perturbations both by planets and by nearby stars would scatter the orbits indiscriminately about the sky, but this would depend in which direction the accretion axis is formed.

The formation process is seen as a continuous one which implies that it could take place whenever the Sun passes through an interstellar cloud. It has been statistically argued that the Sun could encounter at least 3,000 such clouds during its lifetime in the course of the 225 million year revolution round the hub of the Milky Way. Assuming a pessimistic figure of only one hundred encounters, this could generate the order of 200,000-plus comets thought to be populating the solar system at the present time.

Lagrange's ideas of comet origin were based on the idea that they are formed as the result of ejections from the major planets, particularly Jupiter and Saturn. At first the idea attracted few champions. In 1870, the English astronomer Richard Proctor renewed the idea by suggesting that all the short-period comets may have originated from Jupiter. As a giant planet Jupiter has always been somewhat of a mystery, and the area known as the Red Spot has led to considerable speculation as to its nature. Proctor considered that the spot was the site of a super-volcano from which the comets were periodically ejected. Crommelin also subscribed to this belief, and in more modern times the Soviet astronomer Vsekhsvyatsky has argued that there is a very close similarity between the chemistry of cometary gases and those found in Jovian planet atmospheres.

Vsekhsvyatsky has attempted to demonstrate that the shape and characteristics of a cometary orbit depend on the planet from which it originated: the very short-period comets coming

* Approximately a position near the star Sirius (αCMa)

from Jupiter, the longer-period comets from Saturn, Uranus and Neptune and possibly from undiscovered planets lying undetected at remote distances from the Sun. In this way Vsekhsvyatsky equates his speculations with the idea of comet families belonging to different planets (see page 49), an idea which now appears to have little validity.

The most damning evidence against planetary origins is provided by the high ejection velocities necessary to eject a 'comet' away from the surface of the giant planets. In the case of Saturn this would require a velocity of 42 kilometres per second, while for Jupiter – of considerably larger bulk – it would require a velocity of 67 kilometres per second. To overcome this objection Vsekhsvyatsky has switched and developed his ideas to include Jupiter's satellites and possibly even the smaller terrestrial planets as likely sources of ejected volcanic cometary material – particularly that constituting very short-period comets. In the case of Venus or the Earth the ejection velocity requirement is 11 kilometres per second, and for Mars only 5 kilometres.

Nevertheless, in Vsekhsvyatsky's theory there still remains the problem of the short-period orbits, for if short-period comets were ejected from the planets, their present-day orbits retain no evidence of this. The Dutch astronomer van Woerkom (champion of an alternative comet origin idea within the solar system) has concluded that on dynamical grounds the theory simply does not fit the facts. Even if the comets had suffered subsequent perturbations, the distribution of their orbital inclinations is quite wrong. Vsekhsvyatsky considers that the long-period comets are the result of eruptive activity which took place tens or even hundreds of millions of years ago. Because of the sound arguments against the planetary ejection theory, few astronomers outside the Soviet Union now give it more than little credence.

Any comet origin theory to be worthy of serious attention must attempt to account for *all* the physical as well as dynamical properties of comets. Lyttleton and Vsekhsvyatsky both recognize this in their all-embracing ideas. However, one of the most attractive and well-supported ideas for cometary origin within the solar system is one based on dynamical considerations only. This theory argues against the interstellar origin of comets and begins with the assumption that the birth of comets took place some 10^6 to 10^9 years ago through the disintegration of a planet* and that subsequent to the catastrophe the perturbations of Jupiter in particular have driven the majority of the resulting 'comets' further out to beyond the planetary boundaries of the solar system to form a wide circumsolar field.

This origin for comets is linked directly with an idea which also accounts for the origin of the asteroids which lie mainly between the orbits of Mars and Jupiter. An assumption is made that the destroyed hypothetical planet (or planets) consisted of both lighter and heavier elements. The heaviest elements remaining represent the asteroids, and the lighter elements the more fragile comets. The fragments that from the beginning had approximately circular orbits continued as stable members among the interior group of planets and constantly exposed to the intense radiation from the Sun, they soon lost their gaseous envelopes and became small asteroids or ordinary meteorites. But some fragments had elliptical orbits and were therefore immediately subjected to large perturbations by Jupiter and to a lesser extent by other planets. Van Woerkom suggests that the perturbations generally resulted in greatly extending the major axes of the orbits of these fragments, but the exact amount and direction of the perturbing forces depended upon the circumstances of each approach. Van Woerkom shows that the diffusing action of the planets was such that all but 1/30 of the orbits were converted into hyperbolas, and as a result 95 per cent of the exploded material was permanently lost to interstellar space. The theory then argues that since the perturbations cover a continuous range of values, there was an appreciable fraction of comets which were thrown into orbits with major axes ranging in length from 25,000 AU to 200,000 AU. These fragments formed an outer cloud of comets, many of which would have arrived in the cloud within a few years after the explosion and retained much of their gaseous constituents. It is at this point that the theory, as jointly advocated by the Dutch astro-

* Olbers' hypothetical planet which he supposed once lay between the orbits of Mars and Jupiter in the region now occupied by the asteroids (see above).

nomers van Woerkom and Oort, links up with the Fred Whipple 'dirty snowball' idea (see page 59), for they consider these planetary fragments to have solid cores consisting of chunks of iron and stony material (similar in composition to meteorites) embedded in a solid matrix of ice, ammonia, methane and the like.

Oort reasons that once a comet finds itself in the outer cloud, it does not return to the point in space where the explosion took place, since perturbations by passing stars will see to that. It is only when stellar perturbations finally distort the velocity of a distant comet to a considerable extent that it is permitted to return again towards the centre of the solar system, thus reversing the initial process.

The theory argues that from this distant group have originated all comets so far observed including the short-period comets whose orbits are being constantly jostled by the action of Jupiter and the other major planets.

Among alternative ideas of cometary origin, there is one that involves a minimum of conjecture and suggests comets are simply the partially condensed material left over from the original solar nebula much of which, it is supposed, is located at the far boundaries of the solar system. This theory, like the Oort theory, advocates we are able to see these comets through their dynamical interactions with nearby stars, which perturb them inwards towards the Sun. This idea is of interest, since it supposes that comets are remnant 'fossil' material remaining from the original solar nebula – assuming of course that such a nebula did exist and the planets are not captured bodies.

Another theory, suggested by the American T.C. Chamberlin and once widely accepted, is that comets may have originated from the Sun itself. Crommelin also held this view at one time, and he speculated that the Sungrazers might have been born out of the solar prominences thrown up at high velocity from the Sun's surface. The prominences do indeed appear to be ejected solar material travelling at great velocity, but one difficulty with the idea – pointed out by Crommelin himself – is that if the ejection velocity were less than the parabolic velocity of 617 kilometres (383 miles) per second, the prominence material must fall back to the Sun. If it were greater (with hyperbolic velocity), then it would not return to the solar system unless by a remote chance perturbations by Jupiter averted this.

Many short-period comets have orbits very similar to the asteroids. Indeed occasionally it is often difficult to distinguish the two bodies, for frequently a comet will appear quite stellar without indication of a gaseous envelope. Unless it had been previously observed as a comet, such a newly discovered object would automatically be classified as an asteroid unless it possessed a retrograde orbit, and this in itself would make the computers suspicious, since no asteroid so far discovered has retrograde motion.

From the time the first asteroid, Ceres, was discovered by Piazzi in 1801, the question of a possible close relationship between asteroids and comets has often been mooted owing to certain similarities in their eccentric orbits. Nowadays the total population of asteroids is estimated to be greater than 30,000, but most of these are tiny bodies much smaller than Ceres (diameter 700 kilometres). The Englishman Hind startled the astronomical world in 1851 when he announced that Irene, an asteroid he had recently discovered, showed a pronounced nebulous coma. Sir John Herschel confirmed this observation with some of his own, but since then Irene has always remained quite stellar in appearance. Two comets, P/Arend-Rigaux and P/Neujmin (1), have both lacked gaseous envelopes in recent returns; had they been first discovered on these occasions, they would surely have become automatically classified as asteroids.

Öpik has put forward a theory that some of the asteroids of the Apollo group might be representative nuclei of defunct comets – using defunct in the sense that they are now completely degassed bodies, and only an inert nucleus remains. There is a story which illustrates that even dour astronomers on occasions have a sense of humour. When Walter Baade (1892–1960) discovered the minor planet Hidalgo, he said he was undecided whether to call it a minor planet or a comet, but he decided on the former simply because more people were observing minor planets at the time and it would be better taken care of! In a more serious vein it underlines the particular point that it is often extremely difficult to distinguish the two kinds of body simply by their images on the photographic plate.

CHAPTER VI

Famous and Remarkable Comets

Historical Comets

How spectacular were some of the exceptional comets recorded in the past? Were they more brilliant objects than those seen in contemporary times? Neither of these intriguing questions can be answered with any great certainty. Are we to believe, for example, the description left by Diodorus Siculus about a comet which was so brilliant that it cast shadows during the night which supposedly were as intense as those cast by the Moon? Another comet recorded by Seneca shortly after the war of Achaia in 146 BC was described "as large and fiery as the Sun and dispelled the darkness of the nights." Another early description goes perhaps even further, and the comets recorded by Justin, at the time of the birth and accession of Mithridates, were said to have rivalled the Sun in splendour. In March 1402, a brilliant comet seen in Germany and Italy is said to have remained visible in the northern circumpolar regions day and night.

Certainly no comets seen in more recent historical times can rival these early apparitions. Unfortunately, these early apparitions were subject to much exaggeration and cannot be accepted at face value. Nevertheless, it is reasonable to assume that at least one (or perhaps even a number of them) refers to a super-brilliant parent Sungrazer comet which appeared in earlier times and subsequently broke up.

Apart from the early description of Halley's Comet, probably the most reliable extant record of a truly spectacular comet concerns the great Comet of 1264 which displayed a tail more than 100° in length. Contemporaries recorded that when the head of the comet was just clear of the eastern horizon, the tail stretched westwards past mid-heavens. It remained visible to the naked eye for over four months and was widely observed in China and Europe, and both sets of observations agree very closely in spite of the semi-hysterical overtones in European chronicles.

More dispassionate European observations of remarkable comets began after the invention of the telescope in about 1610, and during the seventeenth century a number of brilliant comets were subject to wordy dissertations. Apart from the apparition of Halley's Comet in 1682 the most remarkable was a comet discovered by Godfrey Kirch at Coburg, Saxony, on 14th November 1680 which became a dazzling object with a tail almost 90° long. Edmund Halley also investigated the orbit of this comet and he concluded it was identical with a comet seen in 1106. We now know, however, he was wide of the mark in this particular assumption, for this comet is probably identical with the Sungrazer 1843 I. Whiston, in his book *A New Theory of the Earth*, evolved his fanciful theory of the Deluge based around the 1680 Comet using Halley's erroneous period of 575 years (see also page 138).

During the eighteenth century the most celebrated comet after Halley's was the great comet discovered by Klinkenberg in Haarlem on 9th December 1743 and by De Chéseaux in Lausanne on 13th December. This comet is sometimes incorrectly referred to as De Chéseaux's Comet of 1744. Its head became as bright as Venus, but the most outstanding feature was the magnificent array of multiple tails of which six major divisions were plainly visible spread out like the fan of a peacock's tail, and a total of eleven separate tails were counted during the apparition. On the day the comet arrived at perihelion it was sufficiently bright to be observed with a telescope in broad daylight.

Nineteenth-Century Comets

The first great comet of the nineteenth century, Comet 1811 I, was discovered by Flaugergues of Vivires near Montelmer, on 25th March. It became a circumpolar object in the northern skies, and in the autumn months of 1811 it remained visible throughout the night. This comet was widely observed by the general populace of Europe – more so than many comets before or since. Its apparition gave rise to a whole series of pseudo-phenomena which were supposedly influenced by the presence of the comet in the sky (see also Comet Lore page 144).

The celebrated 1843 Comet (1843 I) is now known to belong to the Sungrazing comet group (see page 90). It was independently discovered by many observers when it appeared as an unmistakable bright glow in the twilight sky early in February 1843. On 28th February it was picked out against the bright background of a daylight sky only 4° from the Sun's edge when its brightness was estimated at between -6^m to -8^m. During March it was a brilliant object in the southern heavens with a tail 70° long. It was not generally seen in Europe until 17th–18th March, but by then it had faded to 0^m with a tail 45° long.

With the next brilliant comet we enter the period of more recent history. On 2nd June 1858, Donati of Florence discovered a telescopic comet moving northwards in the constellation of Leo. As it approached the Sun, it gradually grew brighter and towards the end of August developed a tail which was visible to the naked eye. By September it was at high declination in the northern heavens and hung in the sky like a brilliant scimitar for many hours during the night. After reaching perihelion on 29th September, it began to move back towards the southern hemisphere, and on 5th October the comet passed over the yellow-orange star Arcturus (αBoötes). Practically all contemporary depictions of Donati's Comet show it about this time. These various portrayals disclose that the comet had three well developed tails: two straight gas (plasma) tails and the magnificent scimitar-shaped dust tail (Plate 3).

The next great comet of the nineteenth century was Comet 1861 II, discovered by the Australian sheep farmer John Tebbutt, a self-taught amateur astronomer living in Windsor N.S.W. When Tebbutt found the comet, it was a month before perihelion passage, and it was visible to the naked eye as a 4^m object. Passing from the southern hemisphere into the northern, it created a great sensation as it rapidly increased in brilliance. It became visible in Europe on 29th June, and on

30th June the Earth passed through the comet's tail at a distance of about two-thirds its length from the nucleus.

Sir John Herschel, who during the course of his life had seen many brilliant comets, observed it from his home in Kent and later recalled:

> The comet was noticed here on 29th June by a local resident. His attention was drawn to it being taken by some of his family for the rising Moon . . . when I first observed it – it then far exceeded in brightness any comet I have before observed; those of 1811 and the splendid comet of 1858 not excepted.

In England, on the evening of 30th June, there were numerous reports about the heavens taking on a most unusual appearance. The sky assumed a yellow, auroral-like glare, and the Sun managed to shine through it with only a feeble light. In spite of it being mid-summer, it is related that parish church vicars lit pulpit candles at 7 o'clock, illustrating the sensation of darkness that was truly apparent while the Sun was still some height above the horizon. The comet itself was plainly visible by 7.45 pm.

Although the comet was a brilliant naked-eye object, attempts by the photographer De La Rue to record its impression on two collodion plates were totally unsuccessful, yet many of the background stars near the head were plainly recorded.

Another bright comet, 1862 III, was discovered by Lewis Swift on 15th July. It was not quite as spectacular as some other nineteenth-century comets but was, nevertheless, interesting owing to its faint anomalous tail 30′ long, apparently projected towards the Sun in much the same fashion as that of the bright Arend-Roland Comet in 1957 (see below).

The next great comet was discovered by Jerome Eugene Coggia at Marseilles on 17th April 1874. Unfortunately for northern observers the comet rapidly moved southwards, but before it was lost from view, some remarkable observations were recorded of countless expanding elliptical-shaped envelopes which emanated from the very active nuclear regions in its head. During July 1874 the comet's tail grew to over 40° in length and remained plainly visible long after the head of the comet had set below the horizon.

Three other brilliant comets which appeared towards the latter half of the nineteenth century belonged to the Sungrazer group: 1880 I, 1882 II and 1887 I, and are of such interest that they merit further description (see below).

Twentieth-Century Comets

Among the early comets of the twentieth century, Comet Daniel 1907 IV, discovered on 9th June by Daniel at Princeton, USA, attracted widespread attention. Many photographs were taken which show convincing evidence of tail oscillations with a period of 16 hours.

On 12th January 1910, diamond miners in South Africa's Transvaal were the first to spot the brilliant Comet 1910 I in the morning sky, which later was observed in daylight only 4° away from the Sun. Towards the end of January the tail had grown to a length of 40° and took on a distinctive yellowish tint owing to the characteristic bright sodium emissions present in its head.

The daylight comet was followed by the 1910 apparition of Halley's Comet (see page 33), which was better placed for observation in the southern hemisphere than the northern. On 17th December 1913, Delavan at La Plata in the Argentine discovered Comet 1914 V, which was afterwards visible throughout the world for many months during 1914. This comet was subsequently considered by the superstitious as a cosmic portender to World War I.

During the 1920s and 1930s – apart from Skjellerup's daylight Comet (1927 IX) which was seen only in the southern hemisphere for a few days – there was a dearth of really spectacular comets. In 1943, Comet Whipple-Fedtke-Tevzadze was widely observed in the northern skies, but it was not a brilliant object in the sense of the great comets of the nineteenth century.

In more recent times the first comet to match up to the bright comets of the past was 1947 XII, discovered independently by many observers in the southern hemisphere including the deck watch of a British naval vessel cruising in the South Atlantic. Some reports of the comet particularly noted its marked orange colour, and this was undoubtedly due to the presence of very strong sodium emissions.

Another brilliant comet was seen in 1948 purely by chance,

only 2° away from the limb of the Sun, during a total solar eclipse which occurred at Nairobi in East Africa on 1st November. It later passed into the night sky and developed a tail 30° in length, but by the time the comet had passed into the far northern skies, it had faded below naked-eye visibility.

The year 1957 saw the appearance of two brilliant comets. The first was discovered by Arend and Roland in Uccle, Belgium, on 8th November 1956. During the following months it slowly grew brighter until in April and May 1957 it had become a brilliant naked-eye object in the northern skies, growing a tail over 30° long. Thousands of photographs were taken of the comet, and many of these depicted the remarkable sunward tail 'spike' which appeared to point towards the direction of the Sun when the Earth, during late April, passed through the line of projection of the comet's orbital plane (see Plate 7).

The second comet of 1957 is attributed to the Czech Anton Mrkos – one of the most indefatigable of modern comet hunters – who discovered it on 2nd August with the naked eye while scanning the dawn sky at the Skalnaté Pleso Observatory situated in the Tatra Mountains. It was independently discovered with the naked eye by many other observers in the northern hemisphere, including a British schoolboy, but Mrkos' observation was the first to reach the International Astronomical Union Telegram Bureau then located in Copenhagen.

The Wilson-Hubbard Comet 1961 V became a bright object for a short time. The co-discoverer Wilson was a Pan American airline navigator who sighted it 29,000 feet above the Pacific during the course of an early morning flight from Honolulu to Seattle.* Following this comet, Comet Pereya 1963 V, Comet Ikeya-Seki 1965 VII and Comet White-Ortiz-Bolelli 1970*f* were all Sungrazers.

Apart from the Sungrazers, the next brilliant comet was found on 29th June 1967 by a Queensland, Australia, schoolmaster, Herbert E. Mitchell; it was also independently discovered by another Queenslander, V. Jones on 1st July, and a

* Actually first spotted by a South African Airways stewardess, Miss Anna Ras, during a trans-Sahara flight, but alas, not officially claimed by her in time for her name to be included.

third independent discovery was on 2nd July by Gerber in Argentina, so that the comet eventually became known as Mitchell-Jones-Gerber 1967*f*. It remained a southern object during the short period of its brilliant naked-eye visibility, and a noticeable feature reported more than once was the strong greenish light emitted by the tail.

The year 1969 brought the discovery of two brilliant comets, 1969*g* and 1969*i*. Comet 1969*g* was found independently by a number of Japanese amateurs, and later resolved itself under the title Comet Tago-Sato-Kosaka 1969 IX. Extraterrestrial observations by OAO-2* later revealed a vast hydrogen cloud surrounding this comet. The orbit of Comet Tago-Sato-Kosaka is very eccentric, and in all probability the period is measured in some hundreds of thousands of years.

The second brilliant comet, 1969*i*, was found by John C. Bennet in South Africa on 28th December. During the early part of 1970 it slowly brightened as it travelled northwards in a path almost perpendicular to the orbit of the Earth. For many northern astronomers Bennet's Comet provided the best view ever of a spectacular comet when it became a 0^{m} object in the morning skies towards the end of March 1970. During its apparition it was scanned in hydrogen-alpha light by the OGO-5†, and like Tago-Sato-Kosaka 1969*g*, a vast hydrogen cloud was traced out surrounding the head and tail which measured 13,000,000 kilometres (8,000,000 miles) in the direction parallel with the tail.

Thus during the decade and a half – from the middle 1950s to the early 1970s – several bright comets were observed comparable in brilliance and number to the remarkable comets seen in the nineteenth century. The most remarkable ones, excepting Bennet's Comet, were the Sungrazers, and no less than three of these appeared between 1963 and 1970, repeating the pattern of the three Sungrazers seen during the 1880s.

The Sungrazers

The Sungrazers are among the most remarkable and spectacular comets of all time, and it is not without reason they

* Orbiting Astronomical Observatory 2.

† Orbiting Geophysical Observatory 5.

have earned their colloquial name, for they practically skim the Sun's surface as they pass through the outer, more tenuous atmospheric layers at the time of perihelion passage. These comets are often referred to as the Kreutz group after the Dutchman Kreutz, who in 1888 was among the first to draw attention to the curious similarity in orbits of some of the brightest comets seen in history.

The Sungrazers have been observed in three distinct clusters: the first in the seventeenth century, the second in the nineteenth century and the third which is in progress at the present time. The present series began with Comet 1945 VII (du Toit); then followed 1963 V (Pereya), 1965 VIII (Ikeya-Seki) and 1970*f* (White-Ortiz-Bolelli).

The two strongest clusters are (1) Comets 1880 I, 1882 II, probably the Tewfik Eclipse Comet of 1882 and Comet 1887 I; (2) the four recent Sungrazers 1945–70 (see above).

Within the Sungrazer group as a whole there are two distinct orbital subgroups (see Table III), the existence of which lends tentative support to the idea of a primordial comet (or comets) which broke up in early historical times. In such a break-up only the principal fragments would survive to reappear as independent, spectacular comets.

Even very small differences in orbital velocity of each fragment subsequent to break-up would bring about large differences in their relative orbital positions after only one revolution. In the case of comets originating from a recent break-up, each discrete fragment may be separated by an interval of a decade or more in their respective perihelion passages, while for comets which broke up some thousands of years ago the perihelion intervals for separate fragments might well be a century or more. The smaller fragments resulting from these break-ups would also continue to revolve in similar orbits and give rise to a numerous population of faint 'pygmy' comets visible only from the Earth in special circumstances (see also page 100).

The Ikeya-Seki Comet 1965 VIII was first spotted in the southern hemisphere on 18th September 1965 by two Japanese amateur astronomers, Kaoru Ikeya and Tsutomu Seki, who picked it up, as an 8^m object, quite independently within 15 minutes of each other. There was a great stir of excitement

TABLE III

Sungrazing Comet Group
with Known Orbital Elements (Equinox 1950)

Comet	T (UT)	P(yrs)	e	q	ω °	☊ °	i °
1668	Feb. 28·08		1·0	0·066604	109·81	2·52	144·38
1843 I	Feb. 27·91	512	0·999914	0·005527	82·64	2·83	144·35
1880 I	Jan. 28·12		1·0	0·005494	86·25	7·08	144·66
1882 II	Sept. 17·72	761	0·999907	0·007751	69·59	346·96	142·00
1887 I	Jan. 11·63		1·0	0·009665	58·35	325·50	128·47
1945 VII	Dec. 28·01		1·0	0·006305	50·93	321·69	137·02
1963 V	Aug. 23·92	1111	0·999952	0·005161	85·82	6·77	144·52
1965 VIII	Oct. 21·18	929	0·999918	0·007761	69·03	346·25	141·85
1970 VI	May 18·49		1·0	0·008879	61·30	336·32	139·07

Subgroups with Dates of Perihelion Passage

Subgroup I	*Subgroup II*
–371 B.C.?	1106 Mar. 22?
1668 Mar. 1?	1689 Dec. 2?
1695 Oct. 23?	1702 Feb. 15
1843 Feb. 27	1882 Sept. 17*
1880 Jan. 28	1945 Dec. 27
1882 May 17? (Tewfik)	1965 Oct. 21*†
1887 Jan. 11?	1970 May 18
1963 Aug. 23	

* Probably split-offs originating from same comet on the previous apparition.
† Return of comet 1106?

when the preliminary orbit showed unmistakably that the comet was a Sungrazer which would be ideally placed for viewing both at the time of perihelion and later when it passed into the morning sky. The previous Sungrazers of the twentieth century, 1945 VII and 1963 V, had not been well observed owing to their poor positioning.

The Sungrazer of 1843 (Plate 13) was discovered in broad daylight, and later, when it passed into the night sky, it developed a tail over 70° in apparent length which in terms of true length was equivalent to the distance of Mars from the Sun.

The Sungrazer of 1882 – first comet to be successfully photographed – known at the time as the "September Comet" was discovered by a party of Italian sailors. Later it became a prominent daylight object, and as it grazed by the Sun at perihelion, the nuclear region was seen to split into four bright parts appearing to one observer "like four glistening pearls on a string".

This comet was to register a doubt concerning the chemical make-up of comets which was not to be settled until the Ikeya-Seki 1965 Comet appeared some 80 years later. In 1882, two astronomers, Copeland and Lohse, claimed to have observed the emission spectra of metals and calcium in the head of the comet at the time of perihelion, but others subsequently considered they had been mistaken. In 1965, however, at the Radcliffe Observatory in South Africa and the Haute-Provence Observatory in France, the very same emission lines – due to iron, nickel, chromium, potassium, manganese and silicon, and the H and K lines of ionized calcium – were detected and photographed in Comet Ikeya-Seki, vindicating the long doubted earlier observations.

As it rounded the Sun on the afternoon of 21st October 1965, the Ikeya-Seki Comet was a conspicuous object in the middle latitude skies, but in northern Europe a hazy sky background obliterated all views of the comet. At this time its brightness was estimated at between -10^{m} and -11^{m} or about three magnitudes brighter than its 1843 predecessor. Half an hour before perihelion passage was due to take place, the Japanese astronomer Dr Hirose saw the nuclear region begin to disrupt, and soon two distinct concentrations of luminous material in the head became separated. After passing round the Sun, there was little news about the comet until on 31st October it took the southern hemisphere by surprise as it emerged into the twilight darkness of the morning sky. Professor Bart Bok, then director of the Mount Stromlo Observatory, accidentally noticed it from his bedroom window during an early morning visit to the bathroom and first mistook the 40° tail for a brilliant searchlight beam! Then after realizing his mistake he spent the remaining minutes of darkness before sun-up in unsuccessful attempts to rouse his colleagues and alert the newspapers. Next morning, on 1st November, Australians in their thousands, now forewarned, left their beds *en masse* at 3 am to witness the glorious spectacle that emerged in the morning sky (Plate 12).

Although the comet was also later well observed in Southern Europe and the southern United States, little was seen of it in the northern temperate localities.* Apart from the mass of visual, photographic and spectrographic observations of Ikeya-Seki, several unsuccessful attempts were made to detect the radio wavelengths of the ultraviolet OH emission (represented in the spectrum) in the form of microwave spectral lines of the hydroxyle (OH) radical.

Because of their orbital characteristics, Sungrazers are always better placed for observation in the southern hemisphere than the northern. There can be no doubt that a number of brilliant Sungrazing comets have been missed in the past due to the position of the Earth in its orbit round the Sun during the period mid-May to mid-August. At this time a Sungrazing comet approaching or receding would remain the whole period in the daylight sky. It has been estimated that perhaps one in four Sungrazing comets are missed for this reason. Only by virtue of a solar eclipse did observers see the Sungrazer of 1882 during the few minutes when the Moon totally obscured the Sun. Historical records also mention other comets seen at the time of solar eclipses, and it may be

* In Britain it was only seen by two observers, including the author, who after a three-mornings' vigil, perched on the roof of his house in Buckinghamshire, finally spotted it on the morning of 6th November (using 10×80 binoculars) as it hovered over the sparkling south-eastern horizon like a miniature curl of white smoke.

surmised that many were members of the Sungrazer group.

The recent Sungrazer comet, White-Ortiz-Bolelli 1970*f*, was independently discovered in the southern hemisphere with the naked eye by a number of observers. One of these was Air France pilot Emilio Ortiz during an early morning flight across the Pacific on 21st May 1970.

The Sungrazers are some of the most spectacular bodies observed from Earth and among comets the most significant members of the entire population. The brilliant comet observed in 371 BC and seen by the Greeks to split into two parts, is very likely to have been a Sungrazer. It has been suggested that Sungrazers are the progenitors of all comets, and that other comets observed are split-offs which have subsequently become perturbed into the wide diversity of orbits that we observe today. Some recent work has shown that the Comet 1882 II and the Ikeya-Seki Comet 1965 VIII may be fragments of a brilliant comet which appeared in AD 1106. However, from a study of the diversity of the orbits of the comet population as a whole it is difficult to speculate on how planetary perturbations and differences in velocities resulting from break-up could account for the subsequent wide-ranging distribution of present-day cometary orbits. Out in deep space the Sungrazers revolve in orbits subject to the centre of mass of the solar system; only when nearer the Sun does the Sun itself, as a discrete body, dominate the movement of the comet. As none of the Sungrazers observed pass closer than 3 AU to Jupiter, the influence of this major planet is negligible as a principal agent in their orbital evolution. The fact that the Sungrazers reach perihelion at a point approximately in the direction of the Sun's way in space, is cited as lending some support to the Lyttleton idea of an accretion axis at work (see page 74) or alternatively to some other, at present unrecognized, interstellar capture process.

Encke's Comet

One of the most interesting comets ever observed is the one named after the famous nineteenth-century German astronomer Encke. Its observational history goes back to 17th January 1786, when the French comet hunter Méchain, a great rival of his fellow countryman Messier, spotted a small telescopic comet in the constellation of Aquarius. On this occasion it was only observed twice before becoming hopelessly lost owing to a long succession of overcast skies.

On 17th November 1795, Caroline Herschel – sister of the great Sir William Herschel and in her own right a skilled observer – discovered a comet about 5′ in diameter which was independently found by other observers, and a series of accurate positions were obtained. However, when the traditional parabolic orbit was computed, no sense could be made of it, and the orbital elements remained unresolved.

The story lapses until it is resumed again some ten years later when Pons at Marseilles, Huth at Frankfurt-on-Oder and Bouvard in Paris, all independently discovered a comet on the same night which was faintly visible to the naked eye. When sufficient observations were to hand, the calculation of an orbit was attempted, but the results were inconclusive. Encke now enters the story in the role of computer. After re-examining the data, he found that the observations might after all satisfy a comet with a period of 12.12 years. The comet, after developing a tail 30° long during November finally disappeared and was soon forgotten.

On 26th November 1818, the indefatigable Pons discovered an ill-defined comet which remained visible for nearly seven weeks allowing a long series of positions to be made. Encke quickly acquired them and set about his usual routine task of computing a new orbit using the traditional parabolic method. He was soon made aware that in no circumstances would it be possible to obtain a result. Discarding the old method, he decided instead to try a new computing technique not long developed by K.F. Gauss, a contemporary genius in celestial mechanics whose brilliant new methods had led to the recovery of Ceres (the first asteroid) which had become lost soon after its discovery. With the new method Encke found that the path of the comet could be represented by an ellipse in which the comet revolved round the Sun in the amazingly short time of 3½ years.

Encke's senses were now alerted. He began to search back through the catalogue of comets to see whether the orbital elements resembled those of a previous comet, and was immediately struck by the fact that the comets which had been seen in

1786, 1795 and 1805 might all be the same one. By calculating backwards at approximate $3\frac{1}{2}$-year intervals, his first impressions were soon overwhelmingly confirmed. Next he decided to set about what turned out to be one of the most laborious hand calculations ever performed in the history of celestial mechanics not withstanding the previous efforts of Halley and Clairaut. This meant calculating the effects of planetary perturbations step by step, and for six weeks he laboured night and day almost without stop.* After satisfying himself on the identity of all the past apparitions of the comet, he predicted that it would reappear at perihelion on 24th May 1822, having on this occasion been held back by the perturbing effects of Jupiter for about nine days. History relates that both Arago in France and Olbers in Germany independently arrived at similar conclusions, but they did not extend the work to give it a rigid mathematical proof, so that full absolute credit is rightly attributed to Encke.

The problems involved in confirming the 1822 return was not a simple one. Calculations showed that at this return the comet would not be visible to observers in the northern hemisphere, since its apparent path on this occasion took it far south of the celestial equator. At that time only one observatory was in regular use in the southern hemisphere. This belonged to Governor, Sir Thomas Brisbane at Parramatta, New South Wales, of whom it was said, by his political enemies, was "exceedingly more interested in astronomy than local affairs". Among the observatory staff at that period was Dr Rümker, an assiduous observer who was able to pick up the comet on 2nd June without difficulty, and he succeeded in following it until the 23rd.

This was another great milestone in the history of cometary astronomy, and the greatest since 1759. It was the second discovery of a comet which had been proved to return periodically to the Sun, and the first discovery of a comet with a

* The calculation of perturbations by present-generation computers is a minor exercise in comparison to the months and years of slavery performed in the past by Halley, Clairaut and Encke. For instance, in a recent example a computer wished to calculate the perturbations on a comet at intervals of 10 days in a period extending over 130 years for five separate planets. This took exactly 13 minutes using an IBM 7090, and most of that time was spent sorting out the rectangular coordinates from the magnetic tapes!

very short orbital period comparable to the periods of the terrestrial planets. For his outstanding work, Encke received the Gold Medal of the Royal Astronomical Society in 1824, and the comet became generally known by the name Encke. This decision, history records, was made by Bode* and Olbers. Nevertheless, history also records that Encke himself always modestly referred to it as "the Comet of Pons". Today in the Soviet Union the comet is usually referred to as P/Encke-Backlund. The second name is included in recognition of the later investigations of the comet by O. Backlund.

Since Encke's time the comet has been seen on every apparition during its 3½-year journey round the Sun with the exception of the apparition of 1944 during World War II when it was very unfavourably placed for observation in the southern hemisphere. The comet, owing to its short period, has been the most widely observed of all comets and it has been particularly studied in respect to the problem whether comets show a secular decrease or increase in brightness over a number of revolutions. Although the published results provide contradictory evidence, it is likely that the comet's brightness has diminished very little since it was first observed by Méchain in 1786.

Another very interesting problem about this comet is its so-called secular acceleration in orbit (which has the effect of increasing the orbital velocity and, consequently, bringing about a gradual shortening of the period). Encke first drew attention to this effect in 1838 when he announced that after taking account of every conceivable planetary perturbation, the comet had returned to perihelion on each of its last three apparitions 2½ hours too early. Shortly after Encke's announcement, a suggestion was made that this might be due to the effect of a resisting medium in space, a view widely held for many decades, but it was also suggested that encounters with dense meteor swarms could lead to the same effect. In modern times other comets have been shown to possess similar orbital accelerations and in some instances decelerations. The resisting medium idea has now been discounted, and indeed until as late as the 1960s many of the so-called decelerations and accelerations were attributed by orbit computers to the simple

* Of Bode's Law fame.

fact that comets are inherently difficult objects to measure with precision, and accumulative position errors might suggest comparable effects.

With the introduction of super-fast electronic computers, however, many of these orbits have been rigorously analysed in a manner not possible before, and the conclusion drawn is that there can be little doubt that some kind of non-gravitational force, or forces,* is at work on a number of comets of short period, including perhaps Halley's Comet.

* One suggestion by Whipple is that gas-jet emissions from a solid nucleus may give rise to push-effect non-gravitational forces. Acceleration or deceleration in orbit could then be accounted for by assuming that the nucleus rotates, and the direction of rotation (clockwise or counterclockwise) would determine whether the push-effect would retard the comet or speed it up.

CHAPTER VII

Unusual Comets

Lost Comets

From time to time periodic comets, especially those with short periods, are not recovered at the time of a favourable apparition and are subsequently posted missing. Biela's Comet and Lexell's Comet are classic examples of lost comets, and until 1964, P/Holmes 1892 III was also a member of this group. Some new comets may be observed for one or more apparitions and then fail to reappear when next due. Five main reasons can be listed which may account for a comet's disappearance.

1) Complete disintegration of a comet.
2) A secular decrease in brightness.
3) Comet not suitably placed – in a particular apparition – for observation from the Earth.
4) Insufficient observations obtained in the case of a new comet to calculate a reliable orbit.
5) Severe perturbations on the comet by Jupiter or another planet.

The last one can nowadays be predicted with considerable accuracy. One comet in the past which certainly was whisked away by planetary perturbations is Lexell's Comet. It was discovered by Charles Messier, the famous French eighteenth-century comet hunter, on 14th June 1770. On 1st July it passed very close to the Earth and appeared over 2° 30′ in diameter (or five times the apparent diameter of the Moon). Various orbits were computed, but the first satisfactory one was published by Lexell in St Petersburg and gave the comet a period of just over five and a half years. However, this result was not available until 1778, two years after the comet should have been sighted again. Lexell showed that earlier in May 1767 the comet had been very close to Jupiter. On this occasion, owing to Jupiter's dominance over the Sun in the ratio three to one, the comet was perturbed into its discovery orbit. But Lexell showed that in August 1779 the comet made a further close approach to Jupiter which deflected it once more. The comet was not seen again, and subsequent to Lexell, many other outstanding computers including Burckhardt, Laplace and Le Verrier examined the orbital problems in detail.

Le Verrier studied the perturbing effects caused by the Earth when the comet made its close approach in 1770 which was closer to the Earth than any *known* comet before or since. He found that the comet itself did not have the slightest perturbing effect on the Earth, but the effects of the Earth on the comet were considerable. It seems likely that Lexell's Comet now occupies an orbit beyond Jupiter which renders it invisible from the Earth.

We can almost be certain that other comets will be lost in the future, and new short-periodic comets will be found. The reason for this is that the orbits of periodic comets are being continually transformed by Jupiter's domineering influence (see also page 48). Future perturbations on known comets can be predicted with great accuracy. Comet P/Oterma (page 112) was last observed by Elizabeth Roemer in June and July 1961. It was then subject to Jupiter's influence between July 1962 and January 1964 which transformed it from an 8-year orbital period into one of 22 years, carrying it too far away to be seen from Earth. The comet will be very close to Jupiter again sometime in 2024, but whether this next approach will switch it back into a smaller orbit again, is not certain at the present time. Other comets now invisible from the Earth will sooner or later be subject to Jupiter's influence forcing them into smaller discovery orbits closer to the Sun.

Divided Comets

Although divided, or multiple, comets such as Biela's Comet and some of the Sungrazers are a fairly rare class of object in relation to the total number of comets observed, there are well documented examples going back to earliest historical times.

According to Aristotle, Democritus related the instance of a comet which suddenly divided into a number of stars. Ephorus, a Greek historian, also mentions the division of a comet in 371 BC. Pingré, in his *Cométographie,* records an observation in 1348 about a comet which separated into several fragments, and Kepler suspected that the multiple comets seen in 1618 were originally parts of a single comet. Chinese records provide a wealth of detail about multiple comets, but one of the most remarkable accounts is about the three comets which supposedly *joined* together in 896.

In more modern times the Olinda double comet, discovered by the French astronomer M. Liais at Olinda* in Brazil, on 26th February 1860, is one of the strangest on record. It was not seen by other observers and disappeared two weeks after discovery. Liais' drawing of 27th February shows the larger comet fragment divided into three concentrations of light within the fainter coma, and the nucleus shows two luminous jets directed towards the Sun, followed by a short, broad tail. The smaller fragment consisted of a round nebulosity with marked central condensation. On 10th March, although the luminous jets were no longer visible in the larger fragment, the general appearance had changed little. On 11th March, Liais noted that the larger comet showed a marked tendency to further division, but the following day only a single condensation was seen. Some strong doubts have been expressed about the authenticity of this comet.

The great comet hunter E.E. Barnard told a remarkable story about his observation of multiple comets. It was October 1882 – a month after his second comet discovery – and the great Sungrazing comet of that year was then gracing the morning sky. Barnard related:

> My thoughts must have run strongly on comets during that time, for one night when thoroughly worn out I set my alarm clock and lay down for a short sleep. Possibly it was the noise of the clock that set my wits to work . . . or the worry of the mortgage and the

* It is rare for a comet to take its name from a geographical location. Another example is the Tewfik Sungrazer (see page 88.)

hope of finding another comet or two to wipe it out.* Whatever the cause I had the most wonderful dream. I thought I was looking at the sky which was filled with comets, long tailed and short tailed and with no tails at all . . . I had just begun to gather the crop when the alarm went off and the blessed vision of comets vanished. I took my telescope out in the yard and began sweeping the heavens to the southwest of the great comet in the search for (new) comets. Presently I ran upon a very cometary looking object where there was no known nebula. Looking more carefully I saw several others in the field of view. Moving the telescope about I found that there must have been 10 to 15 comets at this point within the space of a few degrees. Before dawn killed them out I located six or eight of them. That morning I sent a telegram to Dr Lewis Swift, notifying him of the discovery of six or eight 'comets' at a certain position. Whether he thought I was trying to form a comet trust or had suddenly gone demented has never been clear to me, for he unfortunately did not forward the telegram. The observations were amply verified, however, both in this country and in Europe, by other observers who saw some of these bodies. Unquestionably they were a group of small comets or fragments that had been disrupted from the great comet, perhaps when it whirled round the Sun and grazed its surface several weeks earlier with the speed of nearly four hundred miles a second. The association of this dream with the reality has always seemed a strange thing to me.

Since the Olinda Comet 1860 I and the above mentioned Sungrazing Comet 1882 II, a number of other comets have been observed to break up or divide such as 1889 V, 1899 I, 1906 IV, 1915 II, 1915 IV, 1916 I, 1947 XII, 1951 II, 1955 V, 1957 VI and the Ikeya-Seki Sungrazer 1965 VIII. Some of these disruptions have been well photographed and show delicate structural detail that cannot be seen visually.

Various explanations have been put forward to account for double (split or divided) comets. One of the latest suggestions based on the 'snowball' model is that within the nucleus occurs a build-up of heat and pressure owing to heating by trapped radioactive particles. The resulting explosion (or heat shock) shatters the solid matrix into fragments which then become satellite comets, describing independent orbits. Another idea suggests that a disruption of a comet occurs due

* See also page 128.

to its collision with a swarm of meteorites or a small asteroidal body; although statistically this idea seems unlikely, it is perhaps a reasonable assumption and may well be a secondary cause. However, bearing in mind the anomalous behaviour of Comet P/Schwassmann-Wachmann(1) (see page 111), the energy source which causes the splitting appears to lie within the head of the comet itself, but the initial triggering may be coupled with fast-moving corpuscular particles carried into space by the solar wind.

Coloured Comets

Apart from the Sungrazers and other comets which show strong emissions of yellow sodium light when near the Sun, the majority of comets are white – leastwise those we have seen in modern times.

In 49 instances in which colours of comets were recorded by the Chinese, 23 were recorded as white, 20 bluish, 4 red or reddish-yellow and 2 greenish. According to Seneca, the comet of 146 BC was "as large as the Sun" and "its disc a fiery red". Pliny speaks of early comets "whose mane is the colour of blood". According to the French astronomer Arago, the comets of 662 and 1526 were "of a beautiful red".

During the Middle Ages and the Renaissance some very colourful comets were reported, and gold is a colour frequently mentioned, but perhaps the most remarkable are the "blue" comets of 1217 which Pingré cites. Another, in 1476, was "pale blue bordering upon black", a still later comet was described "terrible and of a blackish hue".

Can there by any foundation of truth in these observations? The problem is complicated since many early so-called comets were actually observations of auroral displays, and these occasionally give rise to practically every colour of the rainbow. Another factor is that brilliant comets are often observed near the Sun and therefore are low in the sky at the time of observation. Various meteorological effects can easily influence the colour of a comet. For example, a brilliant sunset certainly will induce a red tint or perhaps even green, and dust clouds held in suspension in the atmosphere following volcanic activity – as with Krakatoa in 1883 – have similar effect. Although one day a genuinely blue or red comet may come along to astound us, on the present available evidence this seems very unlikely, and on this assumption literal colour descriptions of early comets must be treated with great suspicion.*

Eclipse Comets

At the time of total solar eclipse, when the sky is darkened for a period ranging from half a minute to eight minutes, depending on the circumstances of the eclipse, comets, whose existence was previously unknown, have occasionally been seen near the Sun.

In the case of the Sungrazing group of comets, any new member which approaches and leaves the Sun between the period mid-May and mid-August would certainly be missed unless observed at the time of a total solar eclipse. Comets have been positively identified on four occasions during the last 65 eclipses and two on the last 12. However, the number of suspect cometary images obtained on photographs especially exposed for this purpose at the time of total eclipse is considerably greater. In most instances it is difficult to make positive recognitions, since defects in the photographic emulsion cannot be separated from images of small comets unless duplicate plates are exposed.

One of the most famous eclipse comets is the Tewfik Comet, observed at the total solar eclipse which occurred in Egypt in May 1882. This was a magnificent object with a long, distinctive tail and was very likely a Sungrazer. The eclipse comet in 1948 (1st November), observed in the southern hemisphere, was another splendid object, but this particular comet did not belong to the Sungrazer groups.

Probably the earliest extant eclipse comet on record is the one noted by Posidonius (135–51 BC) who relates that during an eclipse of the Sun "a comet became visible which had been hidden through his vicinity". Another early comet first seen during a solar eclipse, on 19th July in AD 418, later passed into the night sky where it was visible for four months.

Many of the fragments of the large Sungrazers such as

* This does not necessarily imply that the ancient observers were barefaced liars. There is good reason to suppose that colour was sometimes invoked to gain more dramatic effect to description, but it was implied in a metaphorical sense rather than a literal one.

observed by Barnard (see above) may still frequent the Sun in large numbers but are much too faint to be seen visually. These may form a class of 'mini or pygmy' comets and may be represented by the faint 'anomalous' objects photographed round the Sun on recent eclipse expeditions.

Comet Holmes

Some of the short-period comets become lost due to perturbations changing their orbits so that they are either ejected completely from the solar system in hyperbolic paths, or they are switched into new orbits lying at greater distances from the Sun where the comet is too faint to be observed from the Earth.

It is considered that many of the 'lost' comets have in fact become lost owing to their complete disintegration as a comet, and such a view was held in the case of Holmes' remarkable short-period comet. This comet was discovered in London in 1892 by the amateur astronomer Edwin Holmes, while engaged in observing the famous Andromeda Nebula with a small reflector in his garden. At the time of discovery it was extremely bright. During the next few weeks it underwent some curious and rapid changes both in brilliance and in form which indicated quite conclusively that here was an unusual kind of comet. When the orbit was computed, it was found that the comet had a period of only 6.9 years. During the 1892 apparition the diameter of the observable coma of the comet was about twice the diameter of the Sun, or 2,400,000 kilometres (1,500,000 miles).

On its next return in 1899 it was again observed, but it never became brighter than 13^{m}. It was also seen on its next return in 1906 as a very faint object at the limit of visibility. In 1912–13 it went missing in spite of intensive searches and was not seen on subsequent returns.

In 1926 the Polish astronomer J. Polak pointed out that perhaps one of the causes of the comet not being seen was that in 1908 it had passed within 80 million kilometres (50 million miles) of Jupiter which resulted in its orbit being changed from a period of 6.9 years to 7.3 years. Nevertheless, even when this was taken into account for the returns of 1928, 1935, 1942 and 1950, it was still not located and was then given up and agreed by most observers that here was a typical case of a comet lost through structural decay.

In 1964 the British computer Marsden, using an IBM 7090 computer, recalculated the motion of the comet from 1899 up to 1975, taking account of every conceivable perturbation. It was found that the comet was theoretically due to arrive at its next perihelion on 15th November 1964 at a distance from the Sun of 2.347 AU. An ephemeris computed by Marsden enabled Elizabeth Roemer to conduct a search with the 40-inch reflector at the United States Naval Observatory at Flagstaff, Arizona, and on 16th April she was able to secure two photographic plates showing tiny but distinct fuzzy images of Comet Holmes shining at $19^{m}.2$. A second pair of plates taken the following night confirmed the discovery beyond all doubt.

The recovery of Comet Holmes in 1964 and again in 1971 on its next return, goes a long way to dispel the widespread belief about the assumed 'death' of short-period comets which have become lost. Nowadays Comet Holmes is much less bright than formerly. In view of the discovery brightness in 1892 it is surprising that such a bright short-period comet had not been found in previous apparitions. This leads directly to the conclusion that in 1892 it is likely that the comet was undergoing an exceptional outburst of some kind, and its normal brightness is more akin to that observed on its recovery in 1964 and subsequent recovery in 1971.

Comet Biela

Over the past 150 years, more than a score of comets have been observed which have shown unmistakable evidence of developing a split, or splits, in their nuclear region. The Sungrazer comets are well-known examples, but probably the most famous split comet is Biela's Comet, which was first seen on 8th March 1772 by Montaigne at Limoges in France. It was seen again when Pons re-discovered it on 10th November 1805, and after a series of observations it was definitely suspected that it may be identical with Montaigne's Comet of 1772.

Nothing further was seen of the comet until it was found independently on 27th February 1826 by an Austrian officer named Biela in Bohemia. On this apparition it was also independently discovered some ten days later by the French astronomer Gambart in Marseilles who, with Clausen, proceeded to compute an orbit when sufficient observations had accumulated. It was found that it fitted a period of $6\frac{3}{4}$ years. The French astronomical community hereafter decided to ignore the claims of Biela and blatantly called it Gambart's Comet. Nevertheless, elsewhere in the world it was known under the more rightful name of Biela.*

The comet was seen again on its return in 1832 but was missed in 1839 owing to unfavourable conditions. On its next return it was first spotted by Di Vico in Rome on 28th November. In December, the English astronomer Hind noted that there was an odd-looking protuberance on one side of the nucleus. On 13th January 1846, Matthew Fontaine Maury† at the US Naval Observatory in Washington detected a double nucleus; two days later Challis at Cambridge in England noticed that the comet was now split into two distinct members.

Poor Challis was at the time much preoccupied with searching for the unknown planet‡ which had been predicted by Adams in England and Le Verrier in France, and at first he made no announcement, for he couldn't quite bring himself to accept that a comet could split in two. In his notes published later he records:

> On the evening of Jan 15, when I first sat down to observe it, I said to my assistant, "I see two comets." However, on altering the focus of the eye glass and letting in a little illumination, the smaller of the two comets appeared to resolve itself into a minute star, with some haze about it. I observed the comet that evening but a short time, being in a hurry to proceed to observations of the new planet . . .

Next night he again saw two comets, and he noted that both comets moved an equal degree retaining their relative posi-

* Strictly speaking it should have been called Montaigne's Comet.

† Better known in his role as the father of oceanography.

‡ Although Challis made the first observation of Neptune, he failed to recognize it as a planet. Credit for discovery went to Galle in Berlin who had a much superior star chart.

tions. He was still reluctant to accept the observation at face value and wrote in his observation book:

> What can be the meaning of this? Are they two independent comets? or is it a binary comet? or does my glass tell a false story? I incline to the opinion that it is a binary or double comet . . . But I never heard of such a thing . . .

The observation was soon confirmed by others, and the double comet was widely observed throughout the world. During March 1846, Maury observed an arc of light extending from the larger fragment to the smaller one, forming a kind of luminous bridge. The larger fragment appeared to develop three tails, each one making an angle of 120° with the next. Slowly the two comets began to separate, until by March 1846 the separation was about 10′. The smaller comet began to grow much fainter in comparison to the larger fragment, and on 15th March the fainter comet was completely lost from view.

The split comet was the astronomical sensation of the decade until displaced by the discovery of Neptune which occurred on 18th September 1846. The comet returned again to perihelion in September 1852 when it was visible for a period of three weeks and provided the remarkable spectacle of one principal comet accompanied by the satellite comet travelling alongside it and separated by a distance of 2.4 million kilometres (1.5 million miles).

In the 1859 return, the conditions for observation were unfavourable, and consequently the comet was not seen. In 1865–66 it was noted that the comet would be ideally placed for observation, and detailed and elaborate searches were carried out. It was known that the comet would approach very close to the Earth on this occasion, and the anticipated spectacle of a bright double comet was awaited eagerly. Nevertheless, it could not be found, and after much fruitless searching everyone finally accepted the idea that the comet had broken up completely, and not a fragment was ever likely to be seen again.

The next approach was due in 1872, and although on this occasion the comet would again theoretically pass close to the Earth, little interest was shown in making further searches.

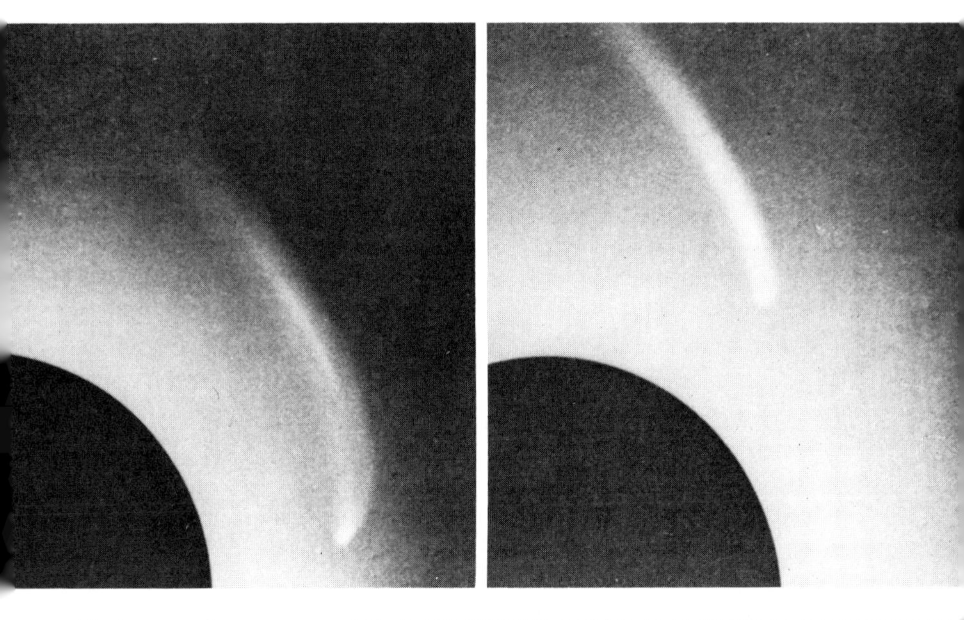

11. Sungrazing Comet Ikeya-Seki (1965 VIII) as it rounded the sun during its perihelion passage on 21st October 1965.

12. Sungrazing Comet Ikeya-Seki as it appeared in the morning sky on 1st November 1965.

13. A contemporary impression of the Great Comet of 1843 as seen over Paris. It developed one of the longest tails on record, which, in terms of true length, was roughly equal to the distance of the orbit of Mars (220 million kilometres or 140 million miles).

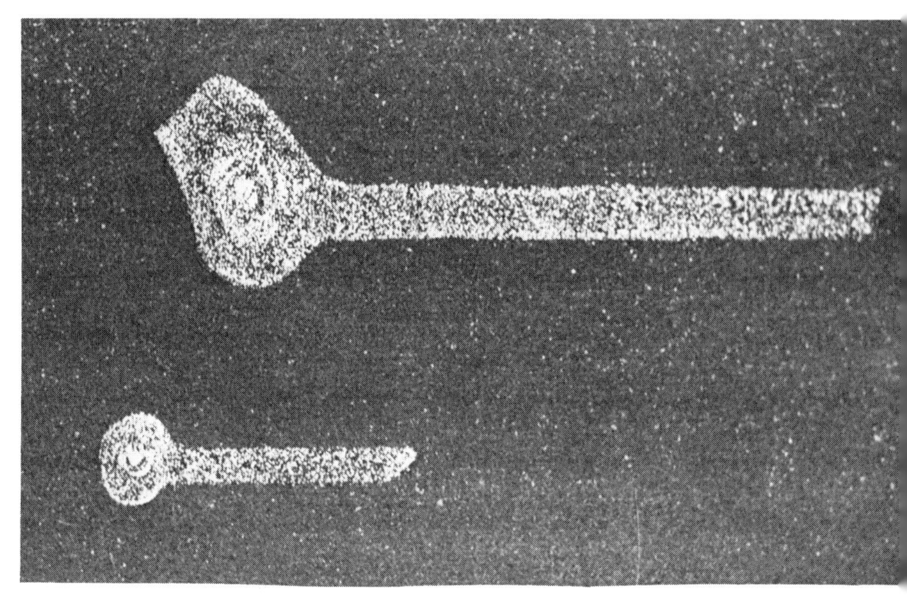

14. The return of Comet Biela in September 1852 showing the remarkable spectacle of one principal comet accompanied by a satellite comet separated by a distance of 2.4 million kilometres.

15. The double Olinda Comet as seen by its discoverer M. Liais in February 1860.

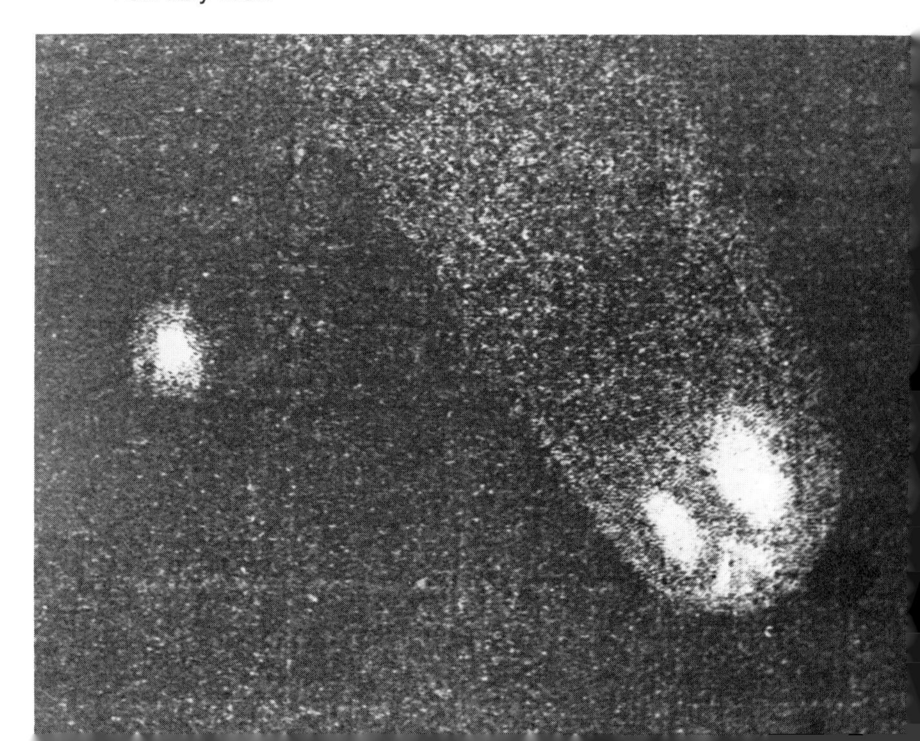

16–17. Comet Morehouse (1908 III) on 29th September 1908 *(left)* and on 3rd October *(below)*.

18—19. Various types of prismatic and Galilean binoculars and field glasses used for amateur comet hunting: *(above)* 10 × 80s, 7 × 35s, 6 × 30s, 8 × 25s, 4 × 30s, 2 × 25s; *(left)* 25 × 105s with 2 × 25s to show comparison in size.

20. Edward E. Barnard as a young man with his 5-inch refracting telescope with which he discovered his early comets.

21. Tycho Brahe, the sixteenth-century astronomer who with his instrumentation and exact observations paved the way for future discoveries.

22. Johannes Kepler, the assistant and successor to Tycho Brahe, known for his laws governing the motions of the planets.

However, late in November 1872 was witnessed one of the most brilliant meteor showers ever to occur (see page 209). The German astronomer Klinkerfues had a sudden inspiration . . . and what followed turned out to be one of the most remarkable episodes in the history of astronomy . . .

After witnessing the meteor shower on the evening of 27th November, Klinkerfues was set to thinking, and he soon came to the firm conclusion that the meteors he had witnessed were in some way associated with the lost Comet Biela. Using the observations of the tracks of the meteors, he traced them back to a point on the celestial globe where they appeared to converge in a radiant. His mind was now working at fever pitch, and making further assumptions, he finally concluded that the lost comet might yet be sighted in the skies of the southern hemisphere.

On 30th November, Klinkerfues, now having fully convinced himself, hastily composed a telegram to the English astronomer Pogson who was then Government astronomer in Madras, India, which read: "Biela touched Earth on 27th: Search near Theta Centauri".

The telegram reached Madras on the night of 30th November by way of Russia in the amazing fast time of 95 minutes.* Pogson later recalled the events after receiving the "startling news":

> I was on the lookout from comet rise (16 hrs local time) to sunrise the next mornings, but clouds and rain disappointed me. On the third attempt, however, I had better luck. Just about 17¼ h mean time, a brief blue space enabled me to find Biela, and though I could only get four comparisons with an anonymous star, it had moved forward 2°.5 in four minutes, and that settled it being the right object. . . .
>
> I recorded it as circular, bright, with a divided nucleus, but no tail and about 45″ in diameter. This was in strong twilight. Next morning, December 3, I got a much better observation of it . . . This time my notes were: circular, diameter 75″, bright nucleus, a faint but distinct tail 8′ in length and spreading . . . I had no time to spare to look for the other comet, and the next morning clouds and rain had returned. For the next three mornings the sky was

* One hundred years later it is doubtful if the 'efficient' present-day telegraph services could even approach it!

quite overcast and afterwards the comet would rise in daylight and could not therefore be observed.

Thus on the face of it Pogson had found the long lost Comet Biela. But had he? In the light of later evidence which is suggested by the orbits of the 1872 meteors and of the probable orbit of the comet, it is likely that the comet was in that part of its orbit no less than 12 weeks earlier, and that any retardation which might have produced so great a delay would anyhow have radically shifted the comet's path.

What seems more likely is that – assuming Pogson had not been misled by a spurious body which is all too easy in comet hunting – his comet discovery was one of pure coincidence, and the comet he saw was totally unrelated to Biela's Comet. However, the mystery has never been satisfactorily resolved, neither has the idea been completely accepted that Biela's Comet broke up after the apparition of 1852.

At first sight perhaps the obvious explanation of the Biela mystery is the appearance of the magnificent display of meteors observed on the evening of 27th November 1872; particularly since we know that the cometary path was very close to the Earth on that occasion. But this assumption is a little too hasty, for the Biela (or Andromedid) meteors had been known long before the comet disappeared. There is evidence that the shower was first seen in AD 524, and there were brilliant displays in 1741, 1798, 1830 and 1838. What is significant, however, is that the Andromedid meteors of 1798,1830 and 1838 originated from a point in their orbit *in front* of the comet's location in this same orbit. Although this evidence provides a strong case for a *genetic* connection between meteors and comets, it provides contra-evidence for the case that meteors are simply cometary *decay* products.

It appears likely that Biela's Comet still exists, but it must be considerably fainter than formerly. With the introduction of fast electronic computers, new orbits have been computed with the idea of providing ephemerides for further intensive searches for their 'lost' comet. The orbit is subject to a rapid regression of the nodes, with the result that the annual meteor shower falls one day earlier every seven years, so that in the 1970s the date is approximately 13th–14th November (see

Appendix). There has been little activity in the Andromedid meteors in recent years, but this does not prohibit a sudden and unexpected brilliant reappearance as occurred with the famous Leonid shower in 1966 (see page 212).

Comet Morehouse 1908 III

A comet discovered by Morehouse at Drake Observatory, Des Moines, Iowa, USA, on 1st September 1908, as a faint 9^m object, and later developed into one of the most unusual comets seen in the twentieth century. The study of the behaviour of its spectacular tail subsequently provided information which led to a better understanding of cometary tail formation.

The comet was ideally placed for study, and during the course of its apparition it travelled from pole to pole, and it remained a circumpolar object in the northern hemisphere for some weeks where it was intensively photographed. (See Plates 16–17).

On 29th September nothing unusual was noted about the tail. The comet from the time of discovery had behaved quite normally, slowly gaining in brightness from 9^m to 6^m. Between 29th September and 2nd October, however, a remarkable change took place. Throughout the night of 30th September the tail changed appearance continuously. By 1st October it had completely disrupted from the head of the comet, and although still visible in photographs, could not be seen visually.

A photograph taken on 2nd October showed three tails. One was a broad fan while the other two were less significant, but all three were changing rapidly. On 15th October a second major change occurred in the main tail which over the past two weeks had grown to a length of 7°. Then suddenly the comet broke in two fragments, but they remained together. A new tail began to form while the old one grew very faint. During November the tail structures became very complex, and long slender rays (plasma tails) emerged from the coma. The principal tail began to show pulsating changes and thereafter repeatedly lost a tail and gained a new one.

For the first time spectrograms taken of the comet revealed the presence of poisonous cyanogen gas hitherto unrecognized in cometary spectra. Such a wealth of photographic reference data was accumulated that even in the 1970s astronomers are still making practical use of it in developing new ideas and models in cometary physics.

Morehouse's Comet was not the first comet to show such unusual activity, but it was better photographed than any before it. During the nineteenth century Brooks' Comet 1893 IV showed similar activity, and after E.E. Barnard photographed it on 21st October 1893, he remarked:

> It presented the comet's tail as no other comet's tail was seen before . . . the tail was shattered. It was bent, distorted, and deflected . . . the whole appearance giving the idea of a torch flickering and streaming irregularly in the wind . . .

Swift's Comet 1892 I also showed unusual behaviour in the tail, and in more recent years Comet Whipple-Fedtke-Tevzadze 1943 I, a prominent object in the northern skies during the early months of 1943, and widely studied by European astronomers in spite of the holocaust raging around them.

Comet Stearns 1927 IV

The fourth comet discovered in 1927 was one of the largest ever observed although paradoxically it never came closer to the Sun to become brighter than $8{\cdot}5^{m}$. It belongs to a group of comets which appear to have *almost* true parabolic orbits which have perihelion distances near to the orbits of Jupiter and Saturn; several other members of this giant group have been observed in recent years. At perihelion, Comet Stearns was 3.6 AU from the Sun, and was so large that it could be followed telescopically over a period of more than four years. When last seen, it was over 11 AU away from the Earth, well beyond the distance of Saturn's orbit. No other comet has ever been observed at such a vast distance.

Other members of this giant group include:

Comet	*Perihelion Distance*
Baade 1955 VI	3.87 AU
Haro-Chavira 1956 I	4.07 AU
Wirtanen 1957 VI	4.45 AU
Abell 1954 V	4.5 AU
Shajn-Comas Solá 1925 VI	4.2 AU
Sarabat 1729	4.05 AU

Comet Schwassmann-Wachmann (1) 1925 II

The first comet jointly discovered by Schwassmann and Wachmann at the Hamburg Bergedorf Observatory on 15th November 1927, and is one of the most remarkable comets that has ever been observed in modern times. It has a planet-like orbit which is almost circular and only slightly inclined to the ecliptic. The comet lies permanently between the orbits of Jupiter and Saturn, and can be observed each year when it comes into opposition with the Earth.

The Schwassmann-Wachmann Comet, however, is not a bright one and can only be followed with certainty using large photographic telescopes during its 16.1 year orbital period round the Sun, during which time it varies in distance between 5.5 AU to 7.3 AU. Normally the comet shines as an 18^m object, but then suddenly and inexplicably it brightens four or five magnitudes, occasionally eight,* within a few days or in some instances within 24 hours. During the brightening the comet also shows marked physical changes within the interval of a few minutes which can be detected visually and photographically. At its minimum brightness it appears as a very faint diffuse comatic cloud, then at the beginning of an outburst it develops a discrete star-like nucleus which is immediately followed by a rapid brightening. The nucleus then proceeds to diffuse outwards, and at the same time the comet begins to diminish in magnitude until the former appearance and brightness is restored. Measurements indicate that during the expansion stage velocities up to 7,000 metres per second are attained. Spectrograms reveal only the presence of a Fraunhofer continuous spectrum in the region of the nucleus, indicating simple reflection; spectrograms of the coma region show banded spectra.

The comet goes on performing its remarkable cyclic outbursts practically every year. One speculative theory put forward to account for this suggests that there is a strong correlation between activity in the comet and terrestrial storms. An analysis of eight outbursts in the period 1939–50 found such a correlation, inferring that ultraviolet solar radiation might be the prime cause, as it might also be with other

* Which implies an increase in brightness by a factor of over 100.

comets which have shown anomalous variations in brightness. But how solar corpuscular radiation may trigger a physico-chemical event is even more speculative. Corpuscular radiation may induce chemical reactions resulting in the generation of explosive material which develops into an ever repeating cycle presumably until the materials responsible in the comet are exhausted. The comet itself is not in a fully stable orbit. Prolonged perturbations by Jupiter are forcing it into a smaller, even more circular orbit, which has resulted in reducing the 16.1 year period into one of only 14 years.

Comet Humason 1962 VIII

Most comets only show tail activity when near the Sun. One notable exception, however, was Comet Humason, a comet discovered during September 1961. It proved to be another example of a giant comet and was a prominent telescopic object even at the distance of Jupiter's orbit. During October 1961, while still 5 AU away from the Sun, it developed prominent gas and dust tails which detached themselves from the head. Spectroscopic observation showed ionized carbon monoxide bands in addition to the Fraunhofer spectrum from reflected sunlight. At perihelion on 10th December 1962, it came within 2.1 AU of the Sun, and later when a definitive orbit was computed, it was attributed with a period of about 2,900 years. No comet previously observed, with the exception of Comet Wirtanen 1957 VI,* had shown such unusual activity at this great distance from the Sun where the influence of solar radiation must be minimal. This lends support to an idea put forward by the German astrophysicist K. Wurm that much of the physical cometary activity is triggered off by internal processes occurring in the comet which are independent of the solar wind corpuscular radiation and the coupled interplanetary magnetic field.

Comet Oterma

In April 1943, Dr Liisi Oterma in Finland discovered a new comet whose orbit was found to be much less elliptical than

* The nucleus of Comet Wirtanen 1957 VI was observed to divide into two parts at about 5 AU from the Sun, and it also showed some unusual tail activity on the sunward side of the head.

most comets'. During the period of revolution of eight years, its distance from the Sun varied between 3.4 AU at perihelion to 4.5 AU at aphelion, so that it remained between the orbits of Mars and Jupiter the whole time in a similar manner to the majority of the asteroids. In fact the comet's orbital elements resembled very closely those of the Hilda group of asteroids of which about 20 members are known. Had the comet not shown a diffuse image on the photograph at the time of discovery, it would surely have been classified as a minor planet, given a number and soon forgotten.

Between July 1962 and January 1964, the comet came into the dominant sphere of Jupiter's influence which for a short time had the remarkable effect of causing the comet to move round the giant planet in the path of an ellipse as might a distant satellite. However, since the comet moves in a direct orbit round the Sun, it evaded capture by Jupiter, and it then moved off into a larger, more eccentric orbit of longer period (about 20 years) which from then on rendered it invisible from the Earth. Comet Lexell, in 1770, is the only other comet whose disappearance can be attributed directly to changes in its orbit resulting from a close encounter with Jupiter.

CHAPTER VIII

Comet Hunting

Comet hunting, as opposed to comet observation, may be said to have begun in earnest when Charles Messier (1730–1817) vigorously applied himself to the task in 1760. Previous to this, Hevelius, Kirch, Klinkenberg, De Chéseaux and possibly Phillip La Hire made some attempts during the seventeenth and eighteenth centuries at regular searches of the heavens with the principal view of finding new comets. Messier, however, was the first to attempt it on a systematic basis and in the following years became so adept and successful that he earned the nickname "the Ferret".*

Messier's actual comet sweeping technique is not known for certain, but from sketchy accounts he left, there is no reason to suppose that it differed much from those in use by present generation amateur-visual observers dedicated to making a persistent search of the skies night after night, month after month and year after year. We know that Messier during the period of his early comet finds used a comparatively small telescope: an inferior 60-millimetre ($2\frac{1}{2}$-inch) refractor giving a magnification of about ×5, which by present-day standards would be considered quite unsuitable. This very point, nevertheless, exemplifies one of the most important factors in comet hunting which is that all kinds of instruments, including the naked eye, have been successfully applied to the task. One cannot be too dogmatic and lay down a strict set of principles which must be adhered to to achieve success.

* Or "bird-nester" of comets according to Delambre who quoted the words used by Louis XV in an eulogy of Messier in 1818.

From a statistical examination of comet discoveries, a number of other significant factors emerge. The majority of new comets are discovered when they are within 2 AU (290 million kilometres, 180 million miles) of the Sun. A detailed examination of the period 1750–1967 reveals that out of a total of 537 comet discoveries only 8 per cent of comets discovered were more distant. During the same period 436 discoveries were visual ones, and 101 discoveries were photographic ones. But in digesting these latter figures it must be borne in mind that the first comet discovered by photography was in 1892. Visual discoveries of comets reach a maximum at elongations from the Sun of about 50°, but the area extends from 15° to 170°. Photographic discoveries tend to be much nearer the meridian (usually near an elongation of 180°, at the time of the observer's midnight).

Having stated the significant statistics, it must be emphasized that comets may be unexpectedly discovered in any part of the sky, at any hour of the night, *or in daylight*, and at any time of the year. But this must be followed by a further qualification: due to the brightness of a comet usually being dependent on its proximity to the Sun, it follows that the brightest comets will be found near the Sun.

In practice there are other determining factors:

1) The age of the Moon (fewer comets are to be found when the Moon is above the horizon).
2) The particular orbital geometry of the Earth-comet-Sun.
3) The geographical location of the observer.

Two other factors are also tied in with 3) above, such as twilight phenomena and extinction (the latter being the visibility of the comet as a function of its altitude above the observer's horizon).

Methods

Statistically it requires anything between 100–200 hours of diligent, premeditated sweeping to discover a comet by traditional visual methods. The averages of the most successful comet hunters in history consistently show the same figures but with occasional large departures from the norm for a particular comet discovery. It may happen by a sheer fluke that a comet is discovered during the first few nights' sweeping, an event which actually occurred when Mark A. Whitaker, a 16-year-old American schoolboy, observing with a 10-centimetre reflector (×45), found Comet 1968*b* on his third night.

Experience, nevertheless, shows that even such lucky discoverers will eventually conform to the norm averages, if they continue to sweep for comets over a period of years.

Visual comet sweepers differ in minor detail as to the precise methods they adopt, but all the consistently successful ones conform to the basic principle of sweeping the sky *slowly* and *methodically*. Experience has shown again and again that the careful examination of a small area of sky in a methodical way nets more discoveries per expended hours than attempts fleetingly to cover the whole sky within the compass of a few nights. An observer's actual technique and methods will depend on the size of his telescope or binoculars, his local sky conditions, his available time, and his temperament. Although there are some notable exceptions, comets are rarely found by observers working together. In order to sustain what amounts to an artistic concentration, it is necessary to adopt an almost Trappist monk-like existence while actually engaged in the business of sweeping.

Although it has already been stated as a broad generalization that comets may be found in any part of the sky, but that the brighter comets tend towards the area surrounding the Sun, it follows that by concentrating on the area above the western horizon after sunset and the eastern horizon before dawn, there is a much greater statistical chance of finding a bright comet here than anywhere else in the sky. The chances are even better in respect to the purely visual observers, since low-altitude twilight zones are not suitable for wide-angle, high-speed camera-telescopes, which therefore concentrate on regions 90°–180° away from the Sun. The western and eastern horizons are particularly suitable for comet hunters using small or "richest field" telescopes and prismatic binoculars. Observers using telescopes in the category 6 to 12-inches may also search these areas, but they are also better suited to regions 45°–180° from the Sun where a comet may be a little less bright and beyond the range of smaller instruments.

The fact that comet hunting is such a time-consuming activity which has no guarantee of success, nowadays places it firmly into the category of amateur-astronomer pursuits. Professional observatories must opt for programmes of work which yield positive results for a given expenditure of time and money. Even in the realm of wide-angle comet photography, few, if any, professional observatories are engaged on definite programmes of comet hunting. Most of the comets found by professional astronomers using photography are simple accidental by-products of some other research programme. A number of observatories, however, devote part of their observing programme schedules to the photographic recovery of predicted periodic comets, many of which are much too faint to be observed visually with even the largest telescopes in use.

The visual observer, if he is to have any chance at all, must learn his sky so that he is on intimate terms with the whole variety of other celestial objects – such as star clusters and nebulae – which during the course of his sweeps may be mistaken for comets. Such knowledge is not gained in a year, and only after several seasons' work will an observer's knowledge be such that when a suspicious object swims into his ken, he will not have to break off observing and resort to his catalogues or star maps to satisfy himself about the true nature of such an object. There will always remain borderline objects, since galaxies and star clusters can be observed in thousands even with moderate-sized telescopes, and the fainter ones are too numerous to be learnt by heart. The brighter ones will be signposts in the sky and so familiar to him that he will know exactly where he is without having to resort to his literary aids. The highly successful Czech comet hunter, Anton Mrkos, who observed during the 1950s, became so adept that he could recognize thousands of galaxies, hazy stars and clusters at a glance and was thus able to spend all his time with his eyes glued to the 'business end' of his 20×100 binocular telescope.

Dark adaptation of the observer's eyes will only come after several minutes of observing and will often go on developing up to half an hour or more after sweeping begins. The eye pupil needs time to dilate in order to reach its peak efficiency as a night vision receptor. Therefore, if catalogues or star atlases need to be consulted during the course of sweeping, it is essential to use only a dim *red* tinted light as a source of illumination so that loss of night vision is kept to a minimum.

When a doubtful nebulous object is detected in the field of view, the only positive identification that the stranger is a comet will be by watching its movement over a period of time. A comet, owing to its orbital movement, will be seen to slowly shift its position against the background stars. Generally speaking an hour or two will readily show this movement. If such a movement is detected, and if it is not a known comet, its position – measured in Right Ascension and Declination by reference to the nearby stars – should be transmitted without delay to the IAU Bureau in Cambridge, Mass., USA via (for preference) the National Observatory of the observer's own country.

Unfortunately spurious comets are also frequently reported, usually by inexperienced observers who do not take the trouble to confirm their discovery before alerting authority.

Mistaken identity may occur sometimes with experienced observers. Some years ago a well-known American astronomer, with one comet discovery already to his credit, was thrilled when sweeping with binoculars he hit upon a large fuzzy object near the horizon. He hurried indoors and asked a companion to come outside and look to the *left* of a star called Kappa Cassiopeia. Yes, his companion could see it, but surely it was to the *right* of the star? Somehow the comet had shifted within a couple of minutes or alternatively it had an unusually large parallax! A more critical evaluation of the fuzzy object proved it to be a lump of damp snow on a telephone wire, dimly illuminated by a street light! Yet another experienced astronomer was fooled in a similar fashion when his 'comet' turned out to be a 'light ghost' reflected off a nearby TV mast.

A few months after the Ikeya-Seki Sungrazer of 1965, a much less cautious professional astronomer thought he had spotted a brilliant comet in the daytime sky and excitedly dispatched a telegram announcing his wonderful discovery to the world. Sad to relate it turned out to be the planet Venus which is often conspicuously visible in daylight. Had he checked beforehand, he would have been saved the subsequent embarrassment of his *faux pas*.

One does not need to be an astronomer to discover a new comet. In 1910 some diamond miners and railway workers in

South Africa discovered the brilliant daylight comet of that year (1910 I) while walking home after finishing a shift. The Comet 1961*d* was found by a professional astronomer named Hubbard but also independently by air navigator A.S. Wilson during a routine early-morning flight. By remarkable coincidence the same comet was independently spotted by an airline hostess as she glanced through the aircraft window (see also page 120). It is sad to relate that the caretaker of the Mount Wilson Observatory also missed being at least joint discoverer of this comet, for he afterwards reported that some days prior to the official discovery he had seen the tail projected above the horizon in the dawn sky.

One of the discoverers of the Comet 1970 was Air France pilot Ortiz who spotted it during a routine airline flight across the Pacific. The Comet 1947 XII was detected in the southern hemisphere by the watch of a British warship then cruising in the South Atlantic Ocean. If one delves further back in history, many of the brilliant comets were first seen by non-astronomers, and there can be no doubt that future bright comets may also be found by such people looking in the right direction at the right time.

Among comet discoveries there are some fascinating stories surrounding particular finds. One of the stories concerns the Dudley Observatory in Albany, New York which had just been rebuilt and re-equipped at great expense by public subscription. Some of the prominent citizens were visiting the Director when one of them remarked that comets were being discovered at other institutions all over the United States, and that Albany was being left behind in the publicity stakes. Whether he was joking or whether he was in earnest, we shall never know, but the Director, Lewis Boss, turned to his assistant and said: "You see, Mr Wells, you must discover a comet." Almost like a fictional hero, Wells did just that within a week! Neither was the comet he found (1882 I) a run-of-the-mill telescopic comet but a fine naked-eye object which remained visible for five months, and which the citizens of Albany could proudly view as their very own.

Among interesting discoveries there are some curious incidents like the one which occurred at the Vienna Observatory on the night 16th–17th November 1890. The astronomer on duty was a 31-year old assistant Rudolf Spitaler, who later became a professor at Prague. About 2.30 am he received a telegram from T. Zona, director of the Palermo Observatory, Sicily, which announced the discovery of a bright comet on the preceding night. With the position of the new comet to hand, Spitaler directed the great 27-inch Vienna refractor at the approximate location in the constellation of Auriga, and on his first glance he caught sight of the comet and then proceeded to measure it as part of a routine observation. He repeated his measurements half an hour later against the known background stars and was surprised that the motion the comet had shown in the interval was less than the figure intimated in Zona's telegram. It was also fainter than he had been led to believe. Suddenly it occurred to Spitaler that he had been observing an entirely different comet which was also unknown. Then slightly shifting the direction of the great telescope, he soon hit upon Zona's much brighter comet a little more than a degree away.

A similar experience also occurred to George van Biesbroeck on 17th November 1925. Setting the 40-inch Yerkes Observatory refractor to observe Comet Orkisz 1925 I, he hit upon a bright 8^{m} comet in the 4-inch finder telescope which had a large field of view. This was later designated 1925 VII and is one of the few comets with a strongly marked hyperbolic orbit.

Amateurs often make accidental discoveries during the course of some other kind of observing programme. In particular variable star observers make a discovery near one of their programme stars. Such was the case with the English astronomer Ryves, observing in Spain on the morning of 10th August 1931, when he spotted Comet 1931 IV near the variable star U Geminorum. A New Zealand observer A. Jones, after abandoning an abortive early morning comet sweep on 6th August 1946 to look at his programme variables, spotted Comet 1946 VI in the field of the variable star U Puppis. The English observer M. Candy found Comet 1961 II while eagerly testing a new telescope which he had poked out of a bedroom window for a quick look at the sky.

Instruments

Up to the time of the invention of the telescope all comets discovered were generally bright naked-eye objects. The first record of a telescopic comet discovery is by Kirch at Coburg who found the Great Comet of 1680 on 14th November that year. Since most of the early telescopes were long-focus instruments and quite unsuitable for comet work, few comets were discovered in the early telescopic era in comparison with optical discoveries in later years. It was some time during the mid-eighteenth century, after the achromatic refracting telescope had been invented by Chester Moor Hall, and the Newtonian reflector had been greatly improved, that telescopic comet discoveries began to increase rapidly.

Messier's earliest telescope, it will be remembered, was a small, short-focus 60-millimetre refractor ×5. Details of his later instruments are difficult to track down. The greatest comet discoverer of all time, Jean Louis Pons (1761–1831), used an instrument he described as the "Grand Chercheur" with a field of view about 3°. No specification is available about the rest of the famous telescope, but it was probably a 5-inch diameter, short-focal-length refractor especially designed for comet seeking. Caroline Herschel (1750–1848) used a beautifully constructed 6-inch Newtonian comet-seeker specially built for her by her famous brother, William, with which she discovered eight comets and several nebulae and star clusters. In the nineteenth century all manner of instruments were used in comet hunting – ranging from Galilean-type opera glasses to the largest telescopes in the world.

It is often assumed that in the choice of visual instruments for comet sweeps the widest possible field of view is necessary in order that the maximum area of sky may be covered in the shortest possible time. In theory the idea appears sound, and indeed many comet-sweepers are designed on the principle of short-focal-length, low-power telescopes providing wide-angle views of the sky. Nowadays such telescopes are referred to as 'richest field' telescopes and may come in the form of short-focal-length *reflectors* (*f*4–*f*5) or *refractors* (*f*8–*f*10). Many low-power binoculars i.e. 7×50s perform the same function. Although these instruments provide splendid wide-angle views of the sky, particularly of large galaxies, nebulae and star clusters, they are only moderate performers in their role as comet-sweepers. The reason is that wide-angle instruments create too bright a sky background, and contrast is an all-important factor in comet sweeping. They also employ magnifications which are too low to resolve small comets, and so objects less than a diameter 2–3′ of arc are easily swept over and missed.

The great nineteenth-century observers, including such names as Barnard, Denning and Brooks, realized that although short-focal-length instruments provided excellent views of *bright* comets and their resplendent tails, the fainter comets were more easily discovered using telescopes of longer focal length. Denning's telescope, a 10-inch reflector, was equipped with powers of ×32, or preferably ×40, which gave true fields of view $1\frac{1}{4}$° and 1° respectively. Brooks used a power of ×40 with his $10\frac{1}{8}$-inch refractor providing a field of 1° 20′, while Barnard, armed with a long-focal-length 5-inch refractor, used a comet eye-piece which provided only a true field of 55′. Yet with this latter instrument Barnard discovered his first eight comets within the space of six years before he started on his career as a professional astronomer.

The nineteenth-century observers of course did not have the opportunity to use the modern giant-sized prismatic binoculars such as the 20×100s, 25×105s (Plate 19) or 25×130s, much in demand today as comet-sweeping telescopes. They all originate as military or naval reconnaissance binoculars, acquired in the surplus equipment markets at a fraction of their true cost. It would be prohibitively expensive for such binoculars to be manufactured for normal peacetime consumption. These instruments are *not* 'richest field' telescopes as some users imagine – although they do provide excellent fields of view ($2\frac{1}{2}$°–3°). The smaller ex-German military 10×80 binoculars (Plate 18) are true 'richest field' telescopes, giving a field of view of over $7\frac{1}{2}$° and providing superb views of the spectacular tails of brilliant comets due to their wide-field abilities. With these latter instruments it is possible to take in truly spectacular views of sprawling star clusters and galaxies, but for the average-sized telescopic comets they do not perform nearly as well as do traditional refracting telescopes of the same aperture.

The great advantage possessed by binocular telescopes is that for periods of extended sweeping they are far more restful to the eyes than the one-eye-open and one-eye-closed method of viewing which is necessary with traditional telescopes. The Czech observers at the Skalnaté Pleso Observatory, employ 20×100 prismatic binoculars designed as German military reconnaissance binoculars for the Eastern Front campaign in World War II. Using this type of binoculars during the period 1946–59, they discovered 17 new comets plus one periodic one. The English school-master, comet sweeper G.E.D. Alcock uses the 25×105s with which he has discovered four new comets and made several independent discoveries of others. Many of the successful Japanese amateur comet hunters also use similar instruments.

Comet Photography

The first really successful photograph of a comet was that of the great Comet of 1882. Later, after Barnard discovered Comet 1892 V by photography – the very first comet to be discovered with such a technique – the number of photographic discoveries rapidly increased.

Nowadays, without exception, the recovery of all expected returns of periodic comets is accomplished by photography. By employing long exposures it is possible to recover comets as faint as 20^m–21^m if one knows exactly where to point the photographic telescope. Many of these faint periodic comets are never seen visually with even the largest telescopes.

Comet discovery by the use of photography – as opposed to (periodic) comet recovery – reached its peak in the 1940–55 after the first wide-angle camera telescopes became operational on routine sky-patrols. During that period at least 50 per cent of all new comets were discovered via their images left on a photographic plate. Subsequent to the middle 1950s, the newer super-Schmidt camera telescopes gradually replaced the older wide-angle instruments. This change-over surprisingly favoured the visual observer again. By employing much shorter exposure times, the images left by a comet are difficult to spot on photographic plates. As a consequence of this limitation, photographic discoveries are often overlooked by casual plate inspection, and the percentage of such discoveries against visual ones dropped from 50 per cent to 30 per cent of the total from the early 1960s onwards.

CHAPTER IX

The Comet Hunters

During the long history of comet hunting there has always existed an atmosphere of keen rivalry among its participants. Occasionally some have become so obsessed with the idea of discovering comets that all else in life becomes secondary. Charles Messier (1730–1817) was just such a man. There is a story that after he had found his twelfth comet and was looking for his thirteenth, his wife fell ill and died. During the last few days of her life, while Messier was busy taking care of her, he was unable to observe, and a comet was discovered by his rival Montaigne of Limoges. After the funeral when somebody offered condolences about his bereavement, Messier was overheard to remark: "Alas, Montaigne has robbed me of my thirteenth comet!" Then suddenly realizing what answer was really expected, he quietly added: "Ah! Poor woman." The story relates that his wife was soon forgotten, and his grieving for the lost comet caused him much anguish and suffering.

Another story about Messier is that on one occasion while he was walking in the President's garden, he was continually gazing up at the night sky on the off chance of spotting a new comet. He became so engrossed that he fell headlong into a well and was laid up in bed for several months.

By a remarkable coincidence the last comet discovered by Messier, the August Comet of 1801, was also the first one to be discovered by Jean Louis Pons (1761–1831). Pons is the most successful discoverer of comets in the whole history of astronomy, and during his time he earned the nickname "the Comet Magnet". What makes Pons's record even more remarkable is the fact that until he was 27 years old he had no astronomical interests. He then became the door-keeper of the Marseilles Observatory and soon began taking lessons from the directors. At the age of 40 he discovered his first comet, and from then on until 1827 it was a rare year indeed that failed to register the discovery of at least one comet by the indefatigable Pons. His grand total of 37 comets is never likely to be beaten or even approached. He once found five comets within eight months during the period February to August 1808, and another five in the twelve-month period August 1826 to August 1827.

In spite of the training given him by Thulis, the Observatory Director, himself a discoverer of comets, Pons was an inaccurate observer. Many of his comet positions are extremely vague, and very imprecise by modern standards. Three of the comets he found in 1808 were so inadequately recorded that no orbits can be calculated for any one of them.

Nevertheless, his fame soon spread through Europe, and in recognition of his success he was promoted in 1813 to the post of Assistant Astronomer at Marseilles. In 1819 he became Director of the Royal Park La Marlia Observatory near Lucca in Tuscany, and during this period discovered seven of his comets. His official title was "Her Majesty's Astronomer Royal", and it is supposed that the Queen offered a cash prize for all new comet discoveries. Pons found a new comet immediately on his arrival, but the early promises by his royal patron were short-lived. Four years later the observatory stipend could not be met. In 1825 he went to Florence under the patronage of Leopold II and was appointed Director of the Observatory and Museum for Physics. In the six years remaining before his death in October 1831 he discovered his last seven comets.

In spite of his zeal as a comet hunter and his desire for self-education, Pons was often treated as a naïve country bumpkin by some of his contemporary astronomers. According to one story – later retold by the influential German astronomer Baron von Zach – Pons, before his first comet discovery, had asked him, as he was a knowledgeable man, the best time to observe for comets. The Baron, seizing the opportunity for a good jest, decided to put tongue-in-cheek and suggested to

Pons that he search diligently for comets when the Sun was covered with spots. Pons gratefully accepted the advice and almost immediately found a comet! Some years later, when asked for the reason of his prolific success as a comet hunter, he related that his searches were always intensified when the Sun was so covered.

It is interesting to speculate further about the story, for it must be remembered that at this period no astronomer had yet discovered the regular 11-year solar cycle when spots increase in frequency and then fall away again to a minimum. Nowadays we know that at time of peak sunspot activity the solar wind also reaches its greatest strength, and we know that the strength of the solar wind has a direct bearing on the size and the activity in a comet's tail. What Baron von Zach suggested as a jest may well have a solid foundation of truth. In more recent times there certainly have been well documented correlations between cometary activity and solar activity to the point when special notifications about intense solar activity are routinely passed on to comet hunters so they may be alerted for possible cometary outbursts.

Among the great comet hunters of the nineteenth century there are ever recurring names in the discovery records such as E.E. Barnard (1857-1923), W.R. Brooks (1844-1921), L. Swift (1820-1913), W.F. Denning (1848-1931), J Tebbutt (1834-1916), J.E. Coggia (1849-1919). F.A.T. Winnecke (1835-1897), W. Tempel (1821-1889), F. Di Vico (1805-1848), H.L. d'Arrest (1822-1875), T. Brorsen (1819-1895), J. Hind (1823-1896), G.B. Donati (1826-1873) and (Dr) H.W. Olbers (1758-1840).

In 1831, the King of Denmark, Frederick VI, instituted a gold medal for all discoveries of new telescopic comets, and the practice was continued until the death of Christian VIII (1848). Some time later the Vienna Academy of Sciences awarded a gold medal to discoverers of new comets, but this practice too was discontinued about 1880. Both these awards greatly stimulated the searches for new comets, but neither of them quite achieved the stimulus to comet discovery as did the cash reward of $200 per comet offered by H.H. Warner, a wealthy American who announced his idea soon after the Vienna Academy's award was terminated.

The award had the effect of involving three men: Barnard, Brooks and Swift into such intense competition in the United States that for several years few comets were discovered by anyone else except in the southern skies. Barnard had discovered his second comet (1881 VI) about this time, and on account of his prize money decided to build for himself and his young wife a new home that might also provide a good view of the sky for further discoveries. He later recalled:

> Times were hard in the last of the seventies and the first of the eighties, and money was scarce. It had taken all that I could save to buy my small telescope. After saving and borrowing and raising a mortgage on the lot, we built a little frame cottage where my mother, my wife, and I went to live. These were happy days, though the struggle for life was a hard one, with working from early to late for the means of a bare existence and the hope of paying off the mortgage, and sitting up all the rest of the twenty-four hours hunting for comets.
>
> We could look forward with dread to the meeting of the notes that must come due. However, the hand of Providence seemed to hover over our heads; for when the first note came due a faint comet was discovered wandering along the outskirts of creation, and the money went to meet the payments. The faithful comet, like the goose that laid the golden egg, conveniently timed its appearance to coincide with the advent of these dreadful notes . . .
>
> And thus it finally came about that this house was built entirely out of comets. This fact goes to prove further the great error of those scientific men who figure that a comet is but a flimsy affair after all; for here was a strong, compact house – albeit a small one – built entirely of them. True, it took several good-sized comets to do it; but it was done nevertheless!

Appropriately enough Barnard's house was named "Comet House". Later, as his fame spread further, he was invited to become an instructor of Practical Astronomy at Vanderbilt University, and eventually he was offered a post as a full-time professional astronomer at the Lick Observatory where he had use of the great 36-inch refractor, then the largest telescope of its kind in the world. When the Yerkes Observatory was established with the even larger 40-inch refractor (which is still the largest of this type in the world), he was offered a new post there where he continued to make some notable discoveries and contributions in numerous fields of observational astronomy. His early training in photography as a youth enabled him to apply this specialized knowledge to astronomical photography then just beginning to be used on a wide-spread scale. The first ever comet discovered by photography (1892 V) was another of Barnard's achievements, and his wide-angle photographs of the Milky Way are classics of their kind even today.

William R. Brooks (1844–1921) discovered comets at the rate of one a year for almost three decades. Like Barnard he began as a self-taught amateur and for much of his observing career worked in comparative isolation and made few contacts among his contemporaries. Although he lived in the United States for most of his life, he was born in Maidstone, England and became interested in astronomy as a boy during a voyage to Australia. His family moved to New York State when he was 13, and the following year he observed Donati's brilliant 1858 Comet through his first home-made telescope.

By profession he was a draftsman and he constructed all his early telescopes himself. He found his first comet in October 1881 (1881 V), as a joint discoverer with W.F. Denning (see below), with a self-made 5-inch *f*10 reflector. This discovery, with the additional impetus of Warner's $200 cash prizes, led him to start comet sweeping in real earnest. He built a $9\frac{1}{4}$-inch reflector and now entered on his life-long task as a dedicated comet hunter to which all else became secondary. In 1888, he was appointed director to a small private observatory erected by a wealthy amateur William Smith who had provided a fine $10\frac{1}{8}$-inch refractor with which his comet hunting was continued.

In 1900, he became professor of astronomy at Hobart College, Geneva, N.Y., USA. His last comet discovery is dated as 20th October 1912. He continued to follow comets up to the day of his death on 3rd May 1921. According to his few friends his death was attributed directly to the physical labours he had endured attempting to photograph P/Pons-Winnecke 1921 III which was recorded by Barnard on 10th April that year.

Brooks's total bag of comets is not easy to assess owing to some confusion with discoveries by others. The comet catalogue of the BAA records a definitive list of 17 different comets in the name of Brooks. Barnard always said that Brooks had discovered 20 comets to his own total of 19. However, careful analysis of Brooks' comet discoveries – including shared ones – reveals a probable total of 25.

William F. Denning (1848–1931) became the greatest British amateur observer of his day. His interests ranged from telescopic observations of Jupiter and comets to naked-eye observations of meteors. He found his first comet as co-discoverer with W.R. Brooks in 1881 and went on to find four more before deciding to concentrate on a long-term, naked-eye meteor observing programme. During the 1880s and 1890s he was a prolific contributor to the observational journals of the day, and he regularly published records of the time he had expended on various projects. Each comet discovery he said had averaged 119 hours of work. His greatest contributions are, however, his catalogues of minor meteor radiants, and he continued observing meteors right up to the time of his death. Due to his great familiarity with the naked-eye stars, he was first to discover Nova Cygni in 1920 when it suddenly exploded into view. An accountant by training, he rejected his professional career and also remained a life-long bachelor in order to devote the maximum possible time to his observational work; during his latter years he lived in extreme poverty and became a neglected, almost forgotten figure in British astronomy.

In spite of the introduction of photography into astronomical observation, amateur astronomers have maintained a dominant position in the field of comet discovery. In the United States, during the early years of the century, Mellish and the Reverend Joel Metcalf accounted for many new comets. Metcalf was also an expert lens grinder and made many of his own telescopes. In 1919 he succeeded in establishing an all-time record by finding two comets within two days, 1919 III and 1919 V. Mellish was greatly encouraged by E.E. Barnard, and during the great astronomer's later years at the Yerkes Observatory, Mellish was allowed the unlimited use of the observatory's comet-seeking telescope.

George van Biesbroeck, the man to close Barnard's eyelids on his death-bed, assumed the role of twentieth-century "Comet Father" after Barnard's death, and he was still observing comets in the early 1970s at the Lunar and

Planetary Observatory in Tucson, Arizona, many years after the authorities of the Yerkes Observatory had considered him to be too old and fragile to be left in charge of telescopes! During his long observing career as a professional he has observed hundreds of comets and encouraged many amateurs to take up the exciting hobby.

It often happens that the enthusiasm and personality of one comet hunter, or comet observer, in a particular locality or country has the effect of triggering off a whole national movement in comet hunting. This first occurred in France during the latter part of the eighteenth century when Messier began his discoveries. He was soon joined by observers such as Montaigne and Méchain, who both reached double figures in their discoveries, and their fame as a trio of comet-hunters-extraordinary must certainly have been an influential spur to Pons to begin his own late start in comet sweeping.

Denning, in England, tried to form an enthusiastic comet sweeping group to rival the formidable 1880s American trio of Barnard, Brooks and L. Swift, but few Englishmen were sufficiently motivated to spend endless long hours outdoors, and what members were recruited soon fell by the wayside.

William Reid, in South Africa, was more successful, and after his first comet discovery in 1918 (1918 II) he assumed – in the 1920s – the role of the southern hemisphere's "Comet Father". Reid wrote and spoke very little about his activities, but in one of his rare public addresses (on the occasion of an award of a medal for his cometary work) he said that the comet which had left the most indelible impression on his young mind was Coggia's great Comet of 1874 (1874 III). Reid discovered six comets on his own account, and his effort at encouraging others was equally successful. There is a story that Reid actually found a comet himself and then telephoned one of his protégés and told him where to look, and afterwards Reid would have no part in claiming the discovery for himself! He greatly encouraged South African observers such as Blathwayt, Ensor and Forbes to make their own discoveries, and he demonstrated what natural advantages exist for comet sweeping in the crystal-clear skies of South Africa.

In the USA, in the 1920s, Leslie Peltier, an Ohio farm boy, began his long career as a successful comet hunter with his discovery of 1925 XI which moved so fast in the heavens it could not be found again for some days. Peltier inherited a famous 4-inch short-focus comet-sweeper previously used by Daniel at Princeton who discovered three earlier comets: 1907 IV, 1909 I and 1909 IV. In latter years, although still an enthusiastic comet observer, Peltier devoted much of his time to variable star observation, and in 1963 he spotted a 3^m nova near the star Vega with the naked eye. His last comet discovery was 1954 XII.

Following World War II, visual comet discoveries were concentrated in two distinct geographical locations. In 1946, at the Skalnaté Pleso Observatory, Czechoslovakia, a long-term programme of comet hunting was begun by professional astronomers using amateur visual sweeping techniques. Ex-German army 20×100 reconnaissance binoculars were chosen as the principal instruments, and in the period 1946–59 (after which the programme was terminated) observers such as Bečvár, Pajdušáková, Mrkos, Kresák and Vozarova made 19 comet discoveries. Although comet observation is still an important part of the observatory's programme, most of it is photographic positional work, and very little comet sweeping is now carried out.

Beginning at the same time as the Czechs, Minoru Honda in Japan began his own long career with the discovery of 1948 IV. Suitable optical telescopes were very difficult to come by in the immediate post-war years in Japan, and when news of his discovery reached the ears of the US Occupying Forces, he was presented with a magnificent pair of 15×100 binoculars. In the early 1960s Japan entered the international comet hunting field in strength with such additional young observers as Seki, Ikeya and Tomita – plus a legion of others who followed a few years later. During the 1960s the Japanese observers almost monopolized the field of visual comet discovery much in the same way as did the Americans in the 1880s.

A good deal of the Japanese success can be attributed to the seriousness with which the Japanese amateur observers approached comet sweeping and to the nationalistic public reaction towards the successful observers. Nowadays, in the West, neither the visual nor the photographic discovery of a new comet stirs much interest outside the close, intimate world of the initiates. The Donohoe medal of the Astronomical Society of the Pacific, which was established in 1890 as a comet award, was discontinued during the early 1950s when many professional astronomers were making numerous accidental discoveries of faint comets using wide-angle Schmidt photographic telescopes. Somehow it did not seem right to award a medal for these accidental discoveries especially when the astronomers concerned were much more interested in their particular field of study usually far removed from comets.

In Japan, during the early 1960s, any successful comet hunter was treated with public adulation. The story of Kaoru Ikeya illustrates this point well. On 2nd January 1963, this 19-year-old amateur – observing from the roof-top perch of his house – spotted his first comet, and next day, when his home was invaded by news photographers and TV and radio interviewers, he became one of the new post-war heroes of Japan.

The story of Kaoru and his comet captured Japanese public imagination. But it was the story of *why* Kaoru Ikeya discovered a comet that held the interest rather than the comet itself. The story begins some years earlier when the business of Kaoru's father began to go downhill, and like many a businessman before him, he took to drinking to forget his troubles. As matters grew worse, more *sake* was consumed, and soon the family faced ignominious disgrace – a tragedy in a country where family pride is almost synonymous with the ultimate purpose of living. His mother was forced to take a full-time cleaning job in a nearby hotel, and Kaoru himself began delivering papers before going to school to earn extra money for the home. Each morning he rose at 5 am to start his morning's work, and during the following months he lived through a period of intense depression. He began to realize that the task of removing the family stigma which had been earned by his father, was going to be his responsibility. But how? Through astronomy? His father, before his business failure, had not favoured such an indulgence by his son. What good was astronomy in business? But Kaoru had ignored this. He was 17 years of age, and the fascination of astronomy held him in its grip. He needed fame, but how could one so young expect to achieve this. It was impossible. No, not quite. If he could attach his dishonoured name to a new comet, his glory and synonymously his family glory would be rung over all Japan.

This idea was, of course, a schoolboy's pipe dream, and not only in Japan have schoolboys dreamt of discovering new comets . . . Leaving school in 1959, Kaoru began an unskilled job in a local piano factory, but his ambition remained as strong as ever. He started to construct a telescope and spent all his free time on the project; after two years' labour he was ready to embark on his nocturnal sweeps of the heavens.

Night after night he continued with unrelenting perseverance, but as the months went by, he began to give way to discouragement. He decided to write to Minoru Honda who with nine comet discoveries to his credit was now famous throughout Japan. Honda wrote back the necessary words of comfort and encouragement, and the search continued . . .

It was 16 months later, and the very day after his mother had prayed in a nearby shrine for good luck, in the year 1963, that Kaoru found his long sought prize. In July 1964, he found his next, Comet 1964*f*, and in 1965 he was joint discoverer with Tsutomu Seki of the great Sungrazing comet which became visible as a brilliant object throughout much of the world.

He became so famous that a film was made about his life, but the producer insisted on using a professional actor to portray the hero, and the romantic story was given a melodramatic twist. Kaoru Ikeya cared little for this doubtful publicity, and said that the simple truth is good enough for him. The richest reward of his life had occurred on the night of 2nd January 1963, when the family honour reappeared in the form of a comet.

In spite of the prolific, almost monopolistic, discoveries by the Japanese observers in the 1960s, the occasional comet was found by other nationals. In England, G.E.D. Alcock, a one time visual meteor observer following the footsteps of Denning, switched over to comet hunting in the middle 1950s, and in 1957, after a long barren period, discovered two comets within a week! In subsequent years he has found two others, plus some independent discoveries of comets found elsewhere. He has also discovered three novae, a feat in itself that places him among the sky-watcher giants of all time.

In the town of Lucas Gonzales, Argentina, lives the Reverend Fredric William Gerber who has one of the largest parishes in the world. After his interest was aroused by the appearance of the Arend-Roland Comet of 1957, he decided to take up comet hunting as a hobby. During his rounds, when he is preaching to his scattered congregations, he is never without his binoculars, and when the sky is clear, he makes his routine scans of the horizon. He was joint discoverer of Comet 1961*e* and 1963*a*, and in 1964, joint discoverer of the Tomita-Gerber-Honda Comet which he spotted with 8×24 binoculars, and in 1967 he discovered the sixth comet of that year using 12×60 binoculars. These discoveries demonstrate conclusively that it is still possible to discover new comets with minimum optical assistance, a lesson indeed to all potential comet hunters.

Among professional astronomers who have been or become dedicated comet watchers of the twentieth century are names such as van Biesbroeck, Jeffers, Cunningham, Whipple and Elizabeth Roemer. Whipple, Cunningham and van Biesbroeck have made comet discoveries. Miss Roemer, a protégé of Cunningham, has been outstanding for her recoveries of periodic comets and her unflagging diligence in photographing faint comets at every opportunity. During the period beginning in the 1940s, numerous comets have been accidentally discovered during the course of other astronomical photographic research programmes by observers with access to large wide-field camera-telescopes. Observers such as Wirtanen, Harrington, Abell and Burnham, whose names constantly recur in discovery statistics of the period, have made many such discoveries.

Comet hunting in the 1970s is still an open field for anyone to join in. Medals and immortal glory are still to be won. The Donohoe Comet Medal (see above) has again been reintroduced, but it is now awarded annually to an amateur astronomer who participates in any significant aspect of cometary work and not necessarily in the discovery of a comet.

The chief requisites for success are a suitable pair of binoculars or a telescope, an intimate knowledge of the night sky, warm clothing and lots of patience and dogged perseverance. The last is the most important of all. The observer must brave the cold night or early morning air, desert family and friends, ignore TV and develop an insouciance to creature comforts and the warm, snug bed which ever waits him indoors. Needless to say such motivated men and women number one in a million.

CHAPTER X

Comet Lore

During the course of history, comets have often been cited as the principal causes of all manner of calamities and bizarre occurrences, ranging in diversity from the Deluge or a particularly excellent wine harvest to the propensity of females to have multiple births.

However, in the light of subsequent events one could hardly blame the inhabitants of the city of London for their ominous concern over comets in the mid-seventeenth century. Daniel Defoe, in his *Journal of The Plague Year* wrote: ". . . a blazing star or comet appeared for several months before the Plague and then did the year after another, a little before the Fire [the brilliant comets of 1664 and 1665]." Of the 1664 Comet he says ". . . it passed directly over London so that it was plain it imported something peculiar to the city alone . . . it was of a faint, dull languid colour, that its motion was very heavy, solemn and slow, and it accordingly foretold a heavy judgement, slow but severe, terrible and frightful, as was the Plague."

Imaginative writers in particular have found comets to be very convenient bodies round which to weave a tale of fantasy, but sometimes it is not clear whether the writers themselves knew the differences between cometary fact and fiction.

It was probably Seneca who first alluded to the Deluge as being caused by a body such as a wandering comet or planet, but it was Whiston (1667–1752), at the beginning of the eighteenth century, who dealt with the question at some length. About the same time the French mathematician Maupertuis

(1698–1759) wrote: "... there are some comets so small that these collisions with the Earth would destroy only few kingdoms, without shattering its mass, but there are others the contact of which might be fatal to every living thing on the globe."

Later writers have invoked comets *ad nauseam* in their stories to account for highly colourful earthly disasters so that in contemporary times the use of comets as a tool of destruction in science-fiction has become a literary cliché. But while perhaps most present-day writers do have a scientific tongue-in-cheek, all the earlier ones were most certainly in deadly earnest. Maupertuis, like Sir William Herschel after him, considered that comets "were peopled by a certain race of men" and that their tails contained "a dazzling train of jewels". Maupertuis, remarking about a possible cometary impact with the Earth, writes in 1742: "... Earth would enjoy rare treasures which a body coming so far would bring to it. We should be much surprised to find that the remains of these bodies we despise are formed of gold or diamonds."

Although Maupertuis' statement sounds extremely fanciful, one should not forget that meteorites on occasion *do* contain diamonds, and some class of meteorite (carbonaceous chondrites) may be associated with comets. However, these diamonds are certainly *not* the brilliant ones worn as costume jewellery which Maupertuis alludes to.

One of the most erudite of the historical pseudo-scientific treatises dealing with an Earth/comet encounter is that of William Whiston. In 1703, Whiston, 25 years younger than Newton, succeeded him in the Lucasian chair of mathematics. In 1693 he had taken deacon's orders and in 1696 published his epoch-making *A New Theory of the Earth* in which he put forward the idea that the geological upheavals recorded in the Book of Genesis and in other scriptures were the result of the Earth's encounter with a comet.

When Whiston formulated his theory, he referred to no comet in particular. However, after Halley had investigated the orbit of the Great Comet of 1680 (as he had successfully done with the 1682 Comet) and announced that it possessed an elliptical orbit giving a period of 575 years, Whiston seized the opportunity of using it for his own ideas, and he proceeded to identify the dates of two of its former apparitions in 2344 BC and 2919 BC, dates previously fixed as times of deluge. Halley's conclusion about the 1680 Comet was later to be proved erroneous by nineteenth-century astronomers, but this was of course not known at the time.

Once Whiston had embarked on his historical comet theory, his imagination knew no bounds. According to his theory, the Earth itself was an ancient comet whose perihelion formerly took it very close to the Sun. These repeated close passages were conveniently used to account for the internal core-heat of the Earth.

In the beginning the terrestrial comet (the Earth) had no rotation, and therefore no alternation between night and day, and as a consequence Whiston maintained it was not able to support living matter. To account for the appearance of life the theory involved itself in assuming the influence of *three* comets. The terrestrial comet had been in existence for a thousand years when it was 'jostled' from its orbit round the Sun by the close encounter with a new comet arrival. The result was that the terrestrial comet was imparted with sufficient motion to cause it to start spinning on its axis.

Life was now able to propagate itself in "multitudinous" forms, and the Earth developed into a kind of paradise about which Whiston waxed eloquently. The third comet (Comet 1680) now enters upon the scene, and so does the theological import of the theory. The 1680 Comet – at a previous apparition – was dispatched by God "to inflict an awful punishment on man for his sins", for by now the human race had progressed to the point when they had got out of hand and become "iniquitous". As a result of the godly command, the tail of the prodigious comet wrapped itself round the Earth and caused the oceans to sweep away and drown the planet's guilty inhabitants "in a glorious religious purge".

Whiston's vivid and detailed account of the catastrophe sets out the events as follows:

> . . . On Friday, 28 November 2349 (BC), or on 2 December 2926 (BC), the comet was situated at its node, and cutting the plane of the Earth's orbit at a point from which our globe was separated by a distance of only 3614 leagues, of twenty-five to a degree. The

conjunction took place at the hour of noon under the meridian of Pekin, where Noah, it appears, was dwelling before the flood.

Whiston then tells of a "prodigeous" tide which split the crust of the Earth. When the tail of the comet "contacted the Earth's atmosphere", the Deluge proper began ("thus were opened up the cataracts of heaven") which resulted in a surge of water "six English miles deep".
Pingré, in a later commentary of Whiston's theory, wrote:

> . . . God had foreseen that man would sin, and that at length his crimes would demand terrible punishment; consequently, he had prepared from the beginning of the creation a comet which he designed to make the instrument of his vengeance.

In hindsight it is difficult to believe that such a theory – even by allowing for the period – could gain an immediate scientific accolade which it certainly achieved when first presented. It was regarded as "the noblest production of genius and science that had ever been given the world". For a time it temporarily eclipsed Newton's *Principia* which had been published only a few years earlier. However, this can be partly explained by the lack of widespread knowledge of Newton's monumental work which used a formal mathematical language only understood by a handful of his contemporaries. Even today there are many who sample the *entrée* and the dessert of the *Principia*, but shirk the highly indigestible main meal of the thesis proper. Whiston's book on the other hand was written in a popular, almost melodramatic, style in a language easily understood by the clergy.

The fact that Whiston had taken deacon's orders in 1693 was highly influential in its wide acceptance. He was a man well known for his strong religious beliefs whereas his scientific contemporaries, such as Newton and Halley, were not. Newton was at one stage accused by Leibniz of atheism, and Halley was accused by Flamsteed as being "a drunken sea captain" when, in 1691, the Savilian Professorship fell vacant, and Halley applied unsuccessfully as a candidate. It is said that Newton fully endorsed Whiston's theory, but as the book was dedicated to Newton, it would certainly have made it awkward had he afterwards decided to revoke its thesis. We do know, however, that secretly Newton did not care very much for Whiston, since it seems he blocked his election to the Royal Society. Halley offered no objection, realizing probably that to do so would put him out of favour again next time a suitable professorship fell vacant.

Whiston was heralded by the clergy for bringing about a much needed accord between astronomy and theology which had been the dream of the eighteenth-century natural philosophers. Nevertheless, within a decade hence, fate caught up with Whiston. In 1710 he was strongly attacked for his beliefs in Arianism, tried for blasphemy and expelled from the University. But the final refutation of his theory did not occur until the end of the eighteenth century when Lexell's Comet passed within 1,200,000 kilometres (750,000 miles) of the Earth without any apparent effects whatsoever on the Earth, whereas the 1680 Comet could never have been closer than 14,400,000 kilometres (9,000,000 miles) at any previous historical apparition.

In modern times a modification of Whiston's ideas (and also those of the Frenchman Freret who later accounted for another historical deluge) has been re-stated by the pseudo-scientific writer Immanuel Velikovsky, who, in 1950, published *Worlds in Collision* which caused a scientific furore. So great was the outcry from informed scientific opinion that the original US publishers, the house of Macmillan, were pressurized into transferring the rights to Doubleday in order to escape the wrath of their bread-and-butter university textbook clientèle who collectively threatened a future boycott of the firm's output.

Velikovsky's book is a startling naïve attempt to discredit the foundation of classical astronomical thought, but it gained some verisimilitude by associating itself with other more acceptable ideas. For example, Velikovsky puts forward the premise that a great comet evolved from an eruption on the planet Jupiter. Now this part is quite in accord with Lagrange's earlier ideas and the mid-nineteenth century ideas of Richard Proctor as well as the contemporary ideas of the Soviet astronomer Vsekhsvyatsky (see page 75). So far so good, but then the same comet, after a series of very unlikely dynamical celestial encounters, finally evolved into the planet Venus! Velikovsky conveniently uses this cometary encounter with the Earth in 1500 BC to cause the Earth to *almost* stop spinning at the time of the Israelite exodus from Egypt. This had the effect of allowing the waters of the Red Sea to divide and enable the Lost Children to cross over the gap into safety.

But Velikovsky, once gripped in the 'comet syndrome', like Whiston and Freret before him, allowed his fertile imagination to probe further. It is perhaps a reflection of the gullibility of our age when a so-called scientific book – running through 15 impressions (1972) and selling many thousands of copies – can put forward for serious consideration the idea that a comet's tail gave rise to a precipitation of carbohydrates in the form of Manna, which supposedly maintained the Israelites for 40 years in the wilderness!

In 1932, another book appeared, innocently entitled: *The Mysterious Comet*, in which the author Comyns Beaumont sets out in typical pseudo-scientific fashion to propound the idea that comets and meteors are the prime cause for all earthly catastrophies such as earthquakes, hurricanes, volcanoes and epidemics. The work of 288 pages is full of ideas more reminiscent of sixteenth-century Ambroise Paré (see page 17)than a twentieth-century writer. Like all self-styled pseudo-scientists, he refutes all accepted ideas about comets without a shred of logical argument to support his own claims. Yet Beaumont, like Velikovsky after him, managed to persuade an academic publisher to underwrite and endorse his nonsensical views.

Edgar Allan Poe, H.G. Wells, Edgar Wallace, Jules Verne and more recently Colin Wilson, among many fiction writers, have all utilized comets as part of a science-fiction backcloth. A majority of science-fiction writers appear to have little real knowledge about cometary phenomena, but leastwise their stories are printed as *fiction.*

In 1829, a certain Mr T. Forster, a scientific writer "of repute", published a work entitled *Atmospheric Cause of Epidemic Diseases* in which he maintained that the most unhealthy periods are those during which some great comet had been seen. David Milne, writing in his *Prize Essay on Comets* in 1830, could scarcely believe that some of his contemporary authors were still equating comets with evil and disaster.

In 1857 a sixpenny pamphlet appeared, bizarrely illustrated, (see Plate 32) entitled *Will The Great Comet now Rapidly*

Approaching Strike The Earth? This referred to the supposed imminent return of the comet of 1264 and 1556. The author, James Bedford PhD, warned the public of the dire consequences of the comet's headlong approach. It is written in a volatile and highly charged pseudo-scientific style which has a comical ring to twentieth-century ears, but it was rapidly digested by a gullible public who bought over 16,000 copies and made it a London street-corner best-seller.

Dr Bedford gives warning that the comet will give rise to a hot season and he goes on to advise: " . . . to keep all kinds of refuse and filth a distance from the dwellings as they would at times of plague; to look well to drainage; to keep no old left off or unused clothes of any kind in holes and corners . . ."

Although this particular comet failed to put in an appearance, when Donati's brilliant Comet appeared in 1858, the lay public mistook it for the long-expected one. As it hung in the sky like a scimitar, Lord Malmesbury wrote in his diary "Everyone now believes in war".

Previous to the expected return of the 1264 Comet, one of the most famous historical panics due to a comet was triggered off in the spring of 1773 when a Parisian rumour spread across France like wildfire.

It all began through the innocent announcement that the French astronomer Lalande was to present a paper before the Academy of Sciences on 21st April entitled "Reflections on the new approach of a Comet to the Earth". Ironically the paper was never actually read on the appointed day simply for lack of time, but the title alone plus a little melodramatic embellishment added by each word-of-mouth as news of it spread across France was sufficient to lead the ignorant peasant population to understand that a comet was about to dissolve the Earth on, or about, 20th–21st May 1773.

When news of its effects reached Lalande, he was horrified and he hastily inserted an announcement in the *Gazette de France* dated 7th May to disclaim any such idea or suggestion on his part. But on 9th May, according to the *Memoirs of de Bachaumont*, it was still widely believed that an enormous deluge threatened, and the archbishop was asked to provide a forty-hour prayer to halt it. The archbishop, however, was dissuaded from doing so by a group of Academicians who persuaded him of the absurdity of the whole affair.

Lalande's paper was finally published in the *Comptes rendus* later in 1773. Its content, sufficient to spark off the earlier panic, was innocuous enough, for it principally consisted of an unambiguous table of figures setting out the distances of the nodes of 61 comets from the Earth's orbit.

Although it was remarked that Milne was surprised by the extent of human gullibility in 1830, events in the twentieth century underline that human nature does not change much. In 1910, when Halley's Comet was at its brightest, itinerant patent medicine vendors did a brisk trade in the United States selling "comet pills" which were guaranteed to ward off any evil influences "that the dreaded star may import". In 1915, when P/Pons-Winnecke reappeared, it portended wartime calamities to many Britishers on the Western Front if pessimistic newspaper accounts of the period are to be believed. When Bennet's Comet was a prominent object in the sky during early 1970, *The Times* correspondent visiting Egypt remarked that the fellahin considered it was some kind of secret Israeli flying object.

Even Halley could not resist the temptation of utilizing a comet to explain the puzzling question of his day of why the climate of North America was more rigorous than that of the same latitude in Europe. Halley supposed that a comet had formerly struck the Earth obliquely and changed the alignment of the Earth's axis of rotation. As a consequence the North Pole, which previously was centred near Hudson Bay, was shifted to a more easterly position, but the vestiges of the Arctic climate had not yet been eliminated from the Hudson Bay area.

The influence of a brilliant comet on a particular wine harvest can be traced back in folk-lore to Roman times. The theory behind the idea is that a brilliant comet has the effect of increasing the atmospheric temperature thus having a profound influence on the alcohol content of the grape. The magnificent claret of 1858 was naturally attributed to the appearance of Donati's bright Comet. The great 1811 Comet also appeared to have been beneficial to both claret and port, and London vintners, even towards the end of the nineteenth century, carried reference in their catalogues to the Comet

23. Isaac Newton, the founder of eighteenth-century physics and mathematics.

24. Edmond Halley, who encouraged and financed the publication of Newton's *Principia* and who calculated the course of the comet named after him.

25. The obsessed eighteenth-century French comet hunter, Charles Messier.

26. Caroline Herschel, who discovered eight comets and several nebulae and star clusters.

27. William F. Denning, the great amateur observer of his day.

28. William E. Brooks, one of the pioneers of celestial photography.

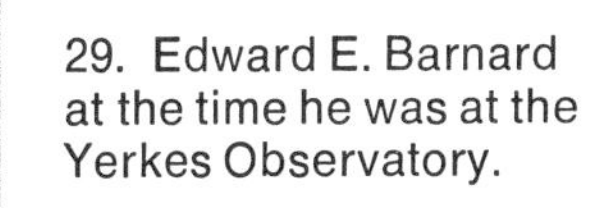

29. Edward E. Barnard at the time he was at the Yerkes Observatory.

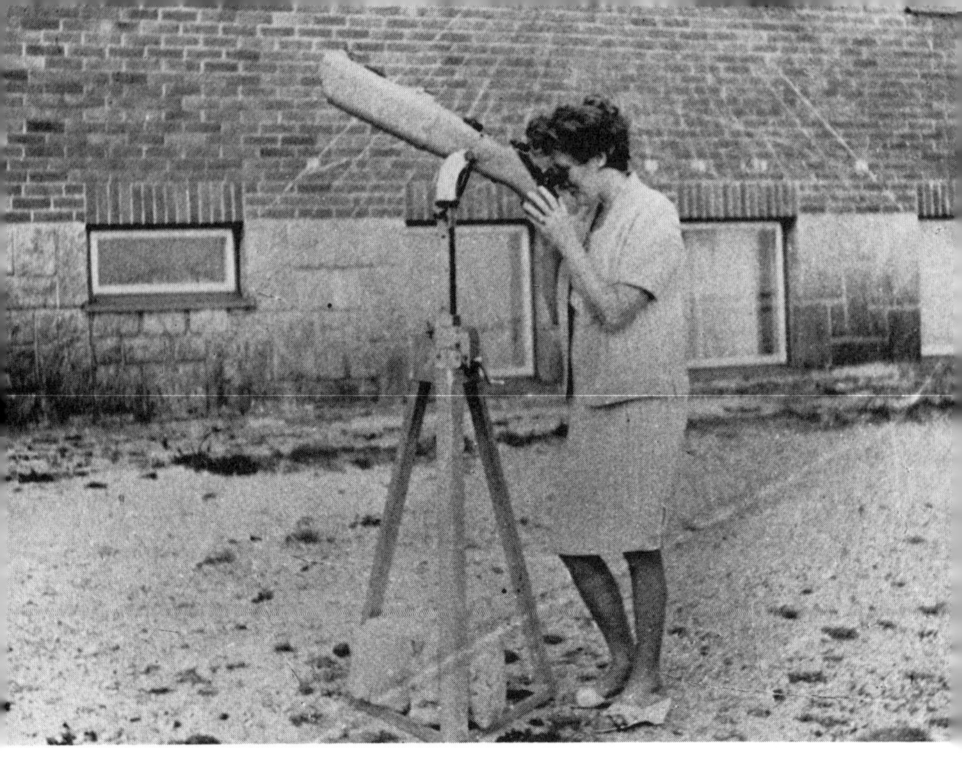

30. Ludmilla Pajdušáková of Skalnaté Pleso Observatory (Czechoslovakia) with the 20 × 100 binocular telescope with which she and her staff discovered 18 comets between 1946 and 1959.

31. Members attending the International Comet Symposium held at Liège in 1965. For key see page 246.

Wine of 1811. This very same comet inspired terror in the Russian peasantry, for Napoleon announced that he looked on it as an omen for his success in the Eastern Campaign. The Little Emperor seems to have been much under the influence of comets. The comet of 1769 – appearing in the year of his birth – was supposedly his protecting genie, and in 1808 Charles Messier actually published a book to uphold the idea which had the splendid title of: *La grande Comète qui apparut à la naissance de Napoléon le Grand.*

One of the most extraordinary stories about the influence of a comet is one concerning the 1680 Comet which tells that in Rome, when it appeared, a hen laid an egg on which was a figure of the comet accompanied by other marks (Plate 34). It was inspected by "the cleverest naturalists" in Rome who remarked "they had never seen such a prodigy before".

Another interesting episode in the annals of comet lore is the celebrated newspaper hoax which appeared in the *San Francisco Examiner* 8th March 1891. That day a story told of a revolutionary automated method of finding comets by photo-electric means. The inventor was none other than the great comet hunter E.E. Barnard who was then working at the nearby Lick Observatory.

When the distinguished astronomer opened his morning paper, his eyes met the following headlines:

DISCOVERS COMETS ALL BY ITSELF

A Wonderful Scientific Invention that will do away with the Astronomers' Weary Hours of Sweeping – The Idea Founded on the Spectrum of the Comet Light – It's Just Like Gunning for Wandering Stars with a Telescope.

Barnard was flabbergasted. Then horrified . . . When he recovered from the initial shock, he read on. But what immediately puzzled Barnard was that the story which followed – although written in a flamboyant journalese style and full of wild speculation – had the suggestion of being concocted by an astronomer or somebody with wide astronomical knowledge. But the article was signed by a Collis H. Burton, a man quite unknown in the astronomical world.

When Barnard reached the part of the news story which purported to quote him verbatim, his eyes almost popped out. After reading a quotation of how the idea supposedly first occurred to him and how it worked, it then continued:

> . . . "Mark now the effect!" cried Barnard, almost rapturously: "When the Moon goes down I will start the telescope 'sweeping'. I can then leave my comet-seeker to its own intelligent work, and give my attention to stellar photography and other important matters. Throughout the night my human telescope explores the skies, stars, nebulae, and clusters innumerable crowd into the field with every advance of the clock, but the telescope gives no sign of their presence, for the analyzing prism spreads out the light of even the brightest among them throughout the length and breadth of the spectrum, and when this spectrum falls on the three slits of the diaphragm its light is far too feeble to exercise any electrical effect upon the selenium!"
>
> "But let even the faintest comet come into range and see what are the consequences! The prism instantly analyses the light, the bright hydrocarbon bands fall upon their respective slits. The light of these, reaching the strip of selenium, so changes the electrical resistance as to disturb the balance of the Wheatstone bridge, and a feeble current is sent through the wire. This in turn closes all the circuits of the powerful Leclanche battery, *and the comet is caught, as in a trap.*"
>
> "An alarm-bell rings in my bedroom down at the cottage. Of course, the signal quickly summons me to the roof. A single glance should suffice to reveal the position of the newcomer."
>
> "Have I tested my invention? Certainly, or I should not speak so confidently. You remember reading the comet discovered by Professor Zona, at Palermo, November 15th of last year? Well, this comet was fairly bright at discovery, but, last month, when my machine was just completed, it had become sufficiently faint to be a most severe test. One night, when the conditions were favorable, I started the finder several degrees from the comet's position, and allowed it to sweep back and forth in the heavens. Sure enough, the distant body – barely visible to the naked eye through the same object glass – was instantly detected, and my experiment proved a complete success."
>
> "You may be sure that I feel pleased: not so much for the honor of the thing (which we all share), but at the immense saving of valuable astronomical time."
>
> By this time I had been so impressed with the grandeur of this invention perfected by these modest workers in astronomical science that I felt impelled to decline their generous offer of further entertainment, and, full of the subject, returned to San

Jose with their entire permission to make the facts known to the public. I am happy to be allowed the honor of communicating to the world this brief sketch of the new invention, which will revolutionize at least one branch of astronomical investigation."

(signed) Collis H. Burton

To say that Barnard was rendered speechless after finishing his news sheet reading would perhaps be an understatement. After recovering his composure, he sat down to write some 'hot' letters of denial to the *Examiner* and to other San Francisco papers. But he had not reckoned on the next twist of events, for the hoaxer had anticipated Barnard's action and had forewarned local newspaper editors that Barnard would vigorously deny the invention after news of it had been published. The upshot was that not one editor would publish a line of his frantic, white-hot disclaimers.

As a result of the article Barnard's mail bag was filled for the next two years. He received letters from all over the world – for the item was internationally syndicated – requesting details and specifications of his wonderful new invention!

Nevertheless, on 5th February 1893 the *Examiner* did finally publish a belated apology shortly after Barnard was in the news again with his discovery of the fifth satellite of Jupiter. Later it was revealed that the hoaxer was an astronomer, Charles B. Hill, an assistant at the Lick Observatory, but it was long suspected that a more famous astronomer, James E. Keeler, was also very much involved. According to Hill they (at the Lick Observatory) were all fascinated by Barnard's uncanny ability to discover faint comets. What began as an innocent leg pull simply to tease Barnard, misfired. The astronomers at Lick had not reckoned that their little scientific joke would receive world-wide and uncritical acceptance by gullible academics.

In its apology to Barnard the *Examiner* "wished him all the new moons and comets that may be necessary to his happiness", but there is no precise record of Barnard's true feelings concerning the hoax. He must have realized, however, that no one meant ill of him, for it was conceived by his colleagues as a gesture of the widespread recognition of his supremacy as a watcher-of-the-skies *par excellence*.

CHAPTER XI

Comet Space Probes

It was remarked in an earlier chapter that it is doubtful whether the true physical nature of comets can ever be determined from earth-based observatories. The orbiting astronomical and geophysical observatories, as for example the NASA OAO-2 and OGO-5, did much to pioneer observations of comets from outside the Earth's atmosphere when the large hydrogen envelopes were detected in comets Tago-Sato-Kosaka 1969 IX and Bennet 1969*i*. These tenuous envelopes were detected by their emissions through the Lyman-alpha radiations which are not able to penetrate the narrow-band electromagnetic windows in the Earth's atmosphere, and thus they remained undetected until the late 1960s when the first orbiting observatories had opportunity to scan a bright comet. No doubt that further advances in cometary physics can be made using the same observatories, but their scope of work is limited by the large distances still remaining between the fixed instrumentation and the comet.

What is required to resolve the outstanding problems concerning the physical nature of comets is a space probe or a series of space probes specifically launched from the Earth to fly-by a comet so as to traverse its coma and study it directly by *in situ* observations. A more ambitious plan would be to attempt a soft landing on a comet nucleus – if such a discrete body is actually found to exist at the centre.

The technical means of accomplishing such an imaginative space mission – leastwise a fly-by near-miss within 10,000 kilometres of the cometary head – has been available since about the middle 1960s. The necessary computed flight times after launching from the Earth to make a comet intercept are less than or (at worst) comparable to the time involved in a space-probe to Mars. But compared to space-vehicle launchings involving planets, comet intercept missions so far have been given low priorities both in the Soviet and American space programmes, since any potential military or tactical benefit appears to be negligible, but the position of priorities may change at any time.

Any space programme involving comet intercept missions needs to plan for at least a minimum of two probes: one to a short-period comet such as P/Encke, and one to a long indeterminate period comet such as a Sungrazer, since there are certainly some differences in physico-chemical characteristics between the two types.

Missions to short-period comets can be planned in detail a long time ahead, since some of these orbits are calculated to a fair degree of precision, and their times of perihelion passage are sufficiently well known to fix the vehicle launch dates from Earth. A study of short-period comets which are due to return to the Sun during the next decade or so between the early 1970s and 1985 shows that only a handful provide favourable launch opportunities. The short list includes P/Encke 1974 return, P/d'Arrest 1976 return, P/Kopff 1983 return and lastly P/Halley in 1985–86. Each of these returns has already been theoretically investigated in relation to a fly-by mission to within 1,000 kilometres of the nuclear region, taking into account acceptable space-vehicle payloads and present-day vehicle guidance systems.

All the intercepts cited above would certainly be practical opportunities, but the orbital velocity of any of these comets is such that at the time of intercept the space-vehicle could not remain close by the comet for very long if present-generation space-vehicles are used. One improvement could be gained by utilizing the gravity of another planet in order to deflect the space-vehicle and impart it with higher velocity such as is conceived with the proposed grand tour of the outer planets in the late 1970s and early 1980s. Alternatively, improvements could be achieved by using electric or atomic propulsion methods at present still in the laboratory/experimental stage.

Missions to long-period comets of indeterminate period offer more difficulties owing to lack of accurate orbital information at the time of launching. These long-period comets would need to be discovered early in their approach to the Sun in order to allow sufficient advance time for the mission to be launched. Such bright comets have appeared in the past as for example Arend-Roland 1957 III and Bennet 1970*i*. Thus the problem of intercept with long-period comets is chiefly concerned with the computation of an accurate orbit at the earliest possible date. However, it may be assumed that a mid-course guidance correction could be made if an improved orbit becomes available during the period of the vehicle's approach flight. Another method of physical guidance might be to evolve an automatic 'comet-seeker' system whereby an onboard spectrometer could lock onto a cometary image by recognizing its characteristic spectral emission.*

Among the short-period comet returns of the near future is Halley's Comet in 1985–86, and this would provide one of the most interesting missions of all time. Unfortunately the Halley mission poses the most difficult problem of all owing to the comet's retrograde orbit round the Sun. This would involve an approach velocity of some 70 kilometres per second as against 0·1 kilometres per second for P/Kopff in 1983. Nevertheless, even in such a brief approach to Halley's Comet, it may be possible to determine the following characteristics:

Comet construction – sandbank or ice-block nucleus?
Diameter.
Definitive chemistry.
Mechanism of gas or dust release or ejection.
Nature of possible internal magnetic forces coupled with the solar wind.

Figure 9 shows a schematic drawing of the launch-geometry involved in the 1985–86 approach of Halley's Comet. Such a probe would need to be launched 210 days ahead of the anticipated fly-by time.

If a probe were made to Comet Kopff in 1983, a rendezvous 'landing' might well succeed there in view of the much smaller

* A not too dissimilar piece of apparatus was the subject of the famous nineteenth century hoax! (see page 145).

approach velocity requirement. Such a rendezvous mission could determine the nature of cometary fragments in great

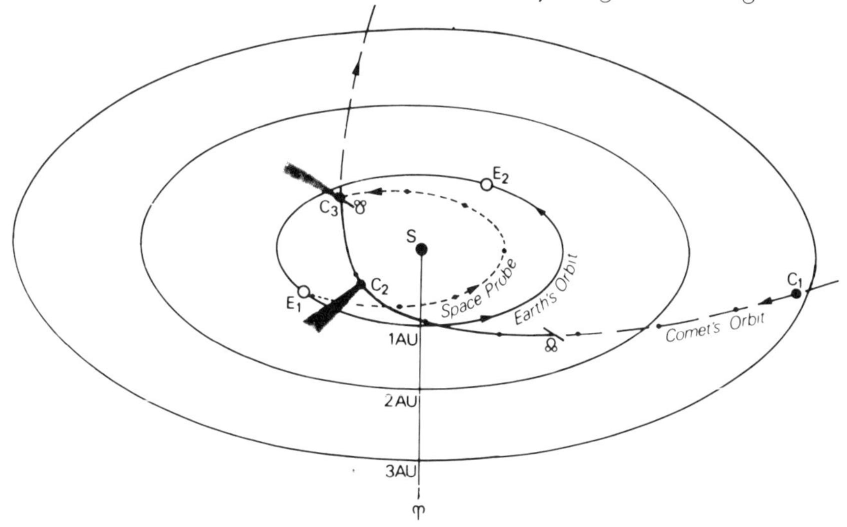

Figure 9. Proposed space probe to Halley's Comet 1985–6.
E_1 = Earth at time of launch 210 days before the space probe and comet intercept.
C_2 = Halley's Comet at time of perihelion passage (approx. 5th Feb 1986).
C_1 = Halley's Comet at time of launch.
E_2 = Earth at time of space probe and comet intercept.
C_3 = Halley's Comet at time of space probe and comet intercept (210 days after launch).
S = Sun.
♈ = First point of Aries.
The intervals between black dots along the comet and space probe orbits represent 30-day periods after launch. See also Appendix II for explanation of comet orbital geometry.

detail. It might also answer the speculations about the possible presence of biological or organic compounds in comets similar to those found in varieties of meteorite which reach the Earth (see also page 222). An understanding of the nature of the comets may yet provide a Rosetta stone to help decipher the enigmatic mysteries surrounding the origin and nature of the solar system as a whole.

CHAPTER XII

Meteorites in History

Up to the beginning of the nineteenth century few in the Western world believed that stones could actually fall from the sky, and most influential scientists of the day dismissed all such reports as nonsense. Although in Western Europe falls of stones had been recorded from the earliest times, all local witnesses to these events were treated with disrespect and ridiculed for their supposed delusions. In 1807 the British Museum had in its possession four "sky stones" which had been presented as curios, but all were treated with the greatest suspicion by the museum's trustees.

One of the oldest of records, although hardly an extant one, is contained in the tenth chapter of the Book of Joshua where it is recorded that during the flight of the Canaanites after the battle of Gibeon, great stones were cast down from heaven so that more were killed by them than with the sword. But these may also refer to a shower of extraordinary large hailstones.*

Diogenes is recorded as believing that meteorites and the stars had a definite connection. Diana of Ephesus stood on a shapeless block which tradition says was a meteoric stone; proof of this is supposedly contained in the speech of the town clerk to placate the riot stirred up against St Paul when he recalled that it fell down in the sky from Jupiter.

Old Chinese manuscripts constantly refer to falls of stones,

* This biblical reference has now unfortunately become a literary cliché. All manner of celestial and terrestrial phenomena have been interpreted from this particular historic event, and it is with some reservation the author includes it here.

and the earliest mention is about 644 BC. A famous stone which most certainly was a meteorite fell in Phrygia and was preserved there for many years. Livy, in his *History of Rome*, described a shower of stones on the Alban Mount about 652 BC, and Plutarch related a similar fall in Thrace about 470 BC, at the time of Pindar. The latter stone, according to Pliny, was still preserved 500 years after the event, but then was mislaid. Pliny also records that a large stony meteorite fell near Egos Potamos in 465 BC which he described as "the size of a wagon and black in colour".

The mass of iron which Achilles offered as a prize at the funeral games of Patroclus – which had been carried off as treasure from the palace of Eetion – was described as crude and *self-fused* and was certainly a meteorite.

A stone which fell in the seventh century is still preserved in Mecca, built into the north-eastern corner of the Kaaba. Another meteorite which gained religious significance fell in Japan during the eighteenth century and for a period was made an annual offering in a temple of Ogi. A stone which fell in India during the nineteenth century was decked with fresh flowers each day and anointed with ghee, and the site of its fall was preserved as a shrine. Some of the oldest recorded iron meteorite fragments form part of a necklace discovered in an Egyptian pyramid and dated to the First Dynasty (3200 BC). In North America, fragments of meteorites, which had been used as tools, have been found in burial mounds of the Hopewell Indians, a race which inhabited the region in prehistoric times. Part of an iron meteorite which fell in 1670 and revered by Tehangin, the great Mongol, was used to forge a ceremonial sword. The oldest authenticated meteorite which is still preserved in a museum is the Ensisheim (Switzerland) stone of 1492, and the local church (translated) records relate the event as follows:

> On the 16th November, 1492, a singular miracle happened: for between 11 and 12 in the forenoon, with a loud crash of thunder and a prolonged noise heard afar off, there fell in the town of Ensisheim a stone weighing 260 pounds. It was seen by a child to strike the ground in a field near the canton called Gisgaud, where it made a hole more than five feet deep. It was taken to the church as being a miraculous object. The noise was heard so distinctly at

> Lucerne, Villing, and many other places that in each of them it was thought that some houses had fallen. King Maximilian, who was then at Ensisheim, had the stone carried to the castle: after breaking off two pieces, one for the Duke Sigismund of Austria and the other for himself, he forbade further damage, and ordered the stone to be suspended in the parish church.

In 1772, three French Academicians, one of whom was the chemist Lavoisier, reported on the analysis of a stone said to have fallen at Lucé on 13th September 1768. This was just about the time that Benjamin Franklin had established the identity of lightning with the electric spark, and the three learned gentlemen were convinced that the Lucé "thunder stones" were figments of fertile peasant imaginations. It does not appear to have entered their heads that a "sky stone" had no relation to a "thunder stone", for the separate events of thunder and lightning and the fall of a meteorite have been confused by mankind since the earliest times, and today the 'visible' terrestrial thunderbolt is still subject to much scientific speculation and controversy. The three French Academicians assured their Academy that there was nothing unusual in the mineralogical characteristics of the Lucé specimen. In their carefully judged opinion it was simply an ordinary stone which had been struck by lightning.

It was not until 1794 that the Czech E.F. Chladni (1756–1827), already well known for his pioneer work on sound waves, compiled an account of numerous reported falls and called attention to the fact that several masses of iron had, in all probability, found their way to Earth from the depths of cosmic space.

One of these irons is still known as the Pallas or Krasnojarsk iron and consists of an irregular mass weighing about 680 kilogrammes (1,500 pounds). It was first described by the German naturalist Peter Simon Pallas (1741–1811), one of a team of scientists commissioned by Empress Catherine (the Great) to explore Siberia (1768–74). It had been found in 1749, lying *in situ*, by a Cossack on a mountain slope between Krasnojarsk and Abakansk in Siberia, and the local Tartars regarded it as a holy thing that had fallen from the sky. A fragment weighing about 3 kilogrammes (7 pounds) was presented by the St Petersburg Imperial Academy of Sciences to the

British Museum where it now forms part of one of the most comprehensive meteorite collections in the world.

Another iron meteorite investigated by Pallas was found in 1783 by local Indians while searching for bees' honey and wax at a spot called Otupa in the Province of Tucuman, Argentina. It was found lying about a foot above the general surface level of the surrounding ground and estimated to weigh more than $1\frac{1}{2}$ metric tons. The local Spaniards at first believed it to be the outcrop of an iron mine, and Don Michael Rubin de Celis was sent by the Viceroy of Rio de Plata to survey and report on it. On arrival in the area the well-trained eye of de Celis soon dismissed the iron-mine theory, for the local geology was all against it. A piece weighing 640 kilogrammes (1,400 pounds) was presented to the British Museum in 1826. This was the principal specimen brought back to Buenos Aires by de Celis; since that time the location of the site has become lost.

Chladni summarized his investigation of the Pallas and Tucuman irons in a book* published in Riga in 1794, in which he presented overwhelming evidence that the two specimens could not have been formed *in situ* as a result of natural geological processes, and that the only alternative and valid conclusion was that they were of extraterrestrial origin and, therefore, must have fallen from the sky. Chladni wrote:

> . . . However, few are willing to believe that in cosmic space, in addition to the larger cosmic bodies, there are many small aggregations of coarse material particles. But this disbelief is an illusory nature and is based not on any theory, but simply upon prejudice . . .

Chladni stated that the flight of such heavy bodies through the sky was the direct cause of the brilliant, luminous atmospheric phenomenon, then known as a fireball. He found few supporters for his ideas, which is a little surprising since even the most sceptical admitted to the existence of the luminous 'meteor' atmospheric events. Chladni's unorthodox views came under heavy fire, and one critic remarked: "He contradicts the entire order of things and does not consider what

* *Über den Ursprung der von Pallas gefundenen und anderer ihr Ähnlicher Eisenmassen, und über einige damit Verbindung* **stehende** *Naturerscheimungen.*

evil he is causing for the moral world."

But for an event which occurred shortly afterwards, Chladni's ideas may have gone unrecognized for a much longer time than they actually did. About 7.00 pm on the evening 16th June 1794, a shower of stones was seen to fall at Siena in Tuscany. A unique account of this event is contained in a contemporary letter:*

> In the midst of a most violent thunderstorm, about a dozen stones of various weights and dimensions fell at the feet of different persons, men, women and children. The stones are of a quality not found in any part of the Sienaese territory; they fell about 18 hours after the enormous eruption of Mount Vesuvius: which circumstance leaves a choice of difficulties in the solution of this extraordinary phenomenon. Either these stones have been generated in this igneous mass of clouds which produced such unusual thunder, or, which is equally incredible, they were thrown from Vesuvius, at a distance of at least 250 miles: judge, then, of its parabola. The philosophers here incline to the first solution. I wish much Sir, to know your sentiments. My first objection was to the fact itself, but of this there are so many eye-witnesses, it seems impossible to withstand their evidence.

In the following year on 13th December 1795, there was another recorded fall. A labourer working near Wold Cottage, a few miles from Scarborough in Yorkshire, was terrified to see a stone fall about ten yards from where he stood. It penetrated 12 inches of soil and 9 inches of solid chalk rock and weighed 25 kilogrammes (56 pounds). No sound or luminous phenomena were noted by the labourer when the stone fell, but in nearby villages there were explosions heard which reminded the villagers of the firing of guns at sea.†

The event was recorded in the *Gentlemen's Magazine*‡ in 1796 where Edward King writes:

* From the Earl of Bristol to Sir William Hamilton then Envoy-Extraordinary at the Court of Naples.

† Perhaps memories of the famous sea battle between the legendary American John Paul Jones and the English Captain Pearson fought off Flamborough Head in 1779.

‡ From an article entitled: "From Remarks Concerning Stones Said to Have Fallen from the Clouds, Both in These Days and in Ancient Times", Volume 66, part 2.

Several persons at Wold Cottage in Yorkshire Dec.13,1795, heard various noises in the air, like pistols, or distant guns at sea, felt two distinct concussions of the earth and heard a hissing noise passing through the air; and a labouring man plainly saw (as we are told) that something was so passing, and beheld a stone, as it seemed at last (about 10 yards, or 30 feet, distant from the ground), descending, and striking into the ground which flew up all about him, and, in falling, sparks of fire seemed to fly from it. Afterwards he went to the place, in common with others who had witnessed part of the phenomenon, and dug the stone up from the place where it was buried about 21 inches deep. It smelled, as is said, very strongly of sulphur when it was dug up, and was even warm, and smoked. It was said to be 30-inches in length, and 28½-inches in breadth, and it weighed 56 pounds.

With the evidence rapidly mounting, it would seem impossible that there could still be considerable doubt about stones falling from the sky. Yet the conservative scientific establishment was still highly sceptical about accepting extra-terrestrial origins for the recovered stones and irons. But Edward King, interpreting the Siena and Wold Cottage falls, says: "... it is suggested that the stones had their origin in the condensation of a cloud of ashes, mixed with pyritical dust and numerous particles of iron, coming from some volcano." King then goes on to develop some involved arguments to show that the Siena stone probably came from Vesuvius while the Wold Cottage stone came from Mount Hecla in Iceland!

In 1798 occurred a well authenticated account of a great fall of stones at Benares in India. No cloud of vapours could be blamed for the occurrence, for none had been seen for a week before the event nor any for a week after. According to the observations of several Europeans and locals, the fall had been preceded by a ball of fire which lasted only a fleeting instant and which was followed by an explosion resembling thunder.

Fragments of these stones plus the stones of Siena and Wold Cottage and another which had fallen at Tabor in Bohemia in 1753 were examined by Edward Howard. On 25th February 1802, he read a paper before the Royal Society in London which reported on the results of chemical and mineralogical examination. Howard concluded:

> The mineralogical description of the Luce stone by the French Academicians of the Ensisheim stone, and the other four places, all exhibit a striking conformity of character common to each of them, and I doubt not but the similarity of component parts, especially of the malleable alloy, together with the near approach of the constituent proportions of the earth contained in each of the four stones, will establish very strong evidence in favour of the assertion that they have fallen on our globe. They have been found at places very remote from each other, and at periods also sufficiently distant. The mineralogists who have examined them agree that they have no resemblance to mineral substances properly so called, nor have they been described by mineralogical authors.

The paper aroused great interest in contemporary scientific circles throughout Europe. But in France the Academicians still remained highly sceptical. A year after Howard's paper, a report came in of yet another fall of stones, and it was as if Nature herself had arranged it that they should fall in the land of the doubting Thomases – France! The matter could now be settled once and for all, and the Minister of the French Interior directed the Academy to send the distinguished physicist Biot to investigate the event on the spot.

The upshot was that Biot, after careful inspection of the stones and cross-examination of the eye-witnesses found:

a) On Tuesday 26th April, 1803 at about 1 pm, there was a violent explosion in the neighbourhood of L'Aigle, in the department of Orne, lasting for five or six minutes which was heard for a distance of 120 kilometres (75 miles).

b) Some moments before the explosion at L'Aigle, a fireball in quick motion was seen from several of the adjoining towns though not from L'Aigle itself.

c) There was absolutely no doubt that on the same day many stones fell in the neighbourhood of L'Aigle.

Biot noted that the stones, which numbered from two to three thousand, had fallen in an area forming an ellipse measuring 10 kilometres (6.2 miles) along its major axis and 4 kilometres (2.5 miles) along its minor axis. Eye-witnesses had been certain that with the exception of a few small clouds of ordinary character, the sky had been completely clear. After Biot presented his report to the Academy, the science of meteoritics had at last become respectable. Chladni's early work, however, seems not to have been mentioned on this occasion, and it was not until several decades later that he was given the respect and recognition long overdue as the true father of meteorite science.

CHAPTER XIII

A Meteorite Falls

Meteorites may fall at any hour of the day, any time of the year and at any geographical location on the Earth. At night, when the sky is clear, the first indication of an approaching meteorite is the appearance of a brilliant, swift-moving *fireball* which is followed shortly after by loud rumblings not unlike thunder.

A fireball differs from an ordinary "shooting star" (meteor or meteoroid) simply by classification. Anything brighter than -4^m, or approximating to that of the planet Venus, is called a fireball, anything less bright is called a "shooting star" (or nowadays more correctly, a *meteoroid*). A brilliant fireball is also sometimes called a *bolide*, from a Greek word meaning 'a thrown spear'. In some definitions a bolide is a brilliant fireball which does not give rise to a solid meteorite subsequently landing on the Earth's surface.

When a meteorite falls in the daytime, it will be accompanied by a sudden intensity of light comparable to the brightness of the Sun and in its wake by a huge dust trail which, in the comparative still air of the upper atmosphere, may persist for several minutes or in exceptional cases several hours.

When a meteorite falls at night, the accompanying brilliant fireball will be seen to change in brightness along the trajectory. Sometimes, towards the end of its flight path it will give the impression of disappearing from view during mid-flight. At this point the meteorite has reached the tropopause zone somewhere about the height of 12 kilometres (7.5 miles) where the retardation, or braking, by the denser part of the Earth's atmosphere has almost brought it to rest. Much of its initial cosmic velocity has now been spent, with the result that the burning stops, and the cooled meteorite falls to the Earth under simple gravitational attraction – its terminal velocity dependent on the air resistance presented to its shape. However, if the meteorite is sufficiently large, or has a high cosmic velocity, it will penetrate the Earth's atmosphere with comparatively little retardation, and such a meteorite will produce a large impact or explosion crater (see below).

After entering the Earth's atmosphere, a meteorite appears like a fast moving star. The exact height at which a meteorite becomes a visible glow is not known, for it is rare that a bolide is seen at the very instance of its entry.

Fireballs are described by observers in a variety of colours such as red, orange, yellow, green, blue, and white. Some of those colour descriptions are real ones, and occasionally a fireball can be observed to run through the whole gamut of colours during its flight across the sky. The colour is a direct measure of the temperature at which the meteorite is burning and a measure of its velocity, hence a white fireball is one travelling faster than a reddish coloured one; the other colours are often representative intermediate velocities when the meteorite is decelerating during its flight path changing from white through to red.

A large fireball may expel several hundred tons of dust into the upper atmosphere, but the daylight 'smoke' trail is the product of both ionized air molecules and solid particles. Such a trail is formed in a perfectly straight or very slightly curved path. The direction of the path is dependent on the position of the observer, and multiple sightings spaced a few miles apart enable the track of the fireball to be accurately plotted, and its pre-entry orbit computed.

The meteorite is only rendered visible from the luminous gas and dust cloud which it produces during its flight. This cloud appears much larger than the object itself and usually takes the form of a round or pear-shaped mass. The apparently large size of the visible phenomenon sometimes leads to wildly exaggerated eye-witness reports about the bolide which produces it. It is not uncommon for an eye-witness to describe a meteorite using dramatic terms like "half-a-mile across".

Later, when the meteoric fragment is located on the ground, it may measure less than a foot. Part of the exaggeration is due to the "over-exposure" dazzle effect which is common to both the human eye and the photographic plate.

The velocity and colour of a meteorite as it passes through the atmosphere is dependent on the previous orbit of the body before it approached the Earth. If the body prior to entry travels a retrograde orbit, such as Halley's Comet, its path will be contrary to the motion of the Earth so that its geocentric velocity might exceed 70 kilometres per second. If it possesses a direct orbit (as does the Earth round the Sun), it may 'overtake' the Earth in its path, and the resultant velocity through the atmosphere may be only a few kilometres per second.

After entry into the Earth's atmosphere, the visible glow begins at a height of approximately 100–120 kilometres (60–75 miles). At this altitude a solid body possessing cosmic velocity begins to collide with air molecules. As the descent continues and the atmospheric density increases, the collisions become so violent and frequent that the outer layer of the meteorite becomes incandescent in much the same way as do space-vehicles during atmospheric re-entries. A temperature of several thousand degrees is reached, and as a consequence of the velocity, an envelope of incandescent gases is trapped and becomes highly compressed in front of the meteorite. Under this pressure a great deal of solid material is liquidized and is stripped off in the form of white-hot luminous droplets which evaporate completely and are dissipated, or they recondense in cooler air to form tiny solid particles.

The luminous 'cloud' surrounding large meteorites during flight may reach a diameter of several hundred meters. But the physics of fireball optical phenomena is not wholly understood contrary to authoritative generalizations usually quoted in popular textbooks. One reason is that it is difficult to simulate identical experimental environments in terrestrial laboratories. Part of the visible light effect is due to the process of luminescence – or cold emission – the result of the collisions of individual air molecules with the meteorite. However, thermal emissions may also play a role – probably more so in the lower regions of the atmosphere. The luminescent trail is caused in part by ionized air molecules, and its luminosity is dependent on the speed of the recombination of the ionized atoms in the upper atmosphere. Occasionally luminous trails have persisted for long periods, and one idea put forward to explain this, suggests that it may be due to the presence of unstable substances in the atmosphere such as nitric oxide which is sometimes observed at high altitudes.

The sound effects accompanying a fall are often more impressive than the visible spectacle, particularly if the fall occurs in daylight hours. The principal cause of the noise is due to the build-up of a compression wave ahead of the meteorite similar to the conical shock wave which develops in front of a supersonic aircraft or a high-speed bullet. The compression wave is the direct cause of a supersonic bang. If the bolide breaks into smaller fragments, as it often does owing to thermal stresses during its descent, each component may trigger off its own set of cannonades and detonations so that the observers on the ground are treated to an impressive and sometimes frightening tattoo of noises. Such sounds can often be heard at more than a hundred kilometres distance from the final impact point; yet, strangely, within the immediate area of the fall silent zones occur and zones of high noise intensification. While the bolide is still high in the atmosphere, many miles from its impact point, peculiar 'singing' or buzzing noises may also be heard. These noises occasionally take the form of a crackling, hissing or rustling. Many observers have described these noises as similar to the eerie effects of wind playing through telegraph wires.

In the past many of the 'singing' and buzzing noises were considered subjective phenomena, but it seems they are real enough and are nowadays wholly accepted. There are instances on record where observers have first been attracted to the passage of a meteorite from the noise rather than the visual appearance of the fireball. When the Irish meteorite of April 1969 passed low over south-east England, it produced a noise like a huge swarm of bees. The loud thunder-like noises originating from the sonic boom travel at normal sound wave velocities, but the other anomalous 'noises' appear to be ultra high-speed phenomena to which the term *electrophonic noises* has been applied; at present time the physics of this phenomenon is little understood.

On the ground, near the point of impact, a wide spectrum of sounds is heard, ranging in diversity from theatrical thunder-claps, express trains and the rumblings of wagon wheels to the occasional staccato cracks not unlike rifle or pistol shots. Animals and humans alike become terrified as buildings tremble and window panes rattle. During the final descent stages of the Barwell (England) meteorite in 1965, an eye-witness likened it to a dive-bomber attack he had experienced in World War II. This person was badly shaken and suffered considerable physical shock in the period following the fall. In this particular fall, the author interviewed an eye-witness who noted that in a nearby field some horses were seen to panic and bolt seconds before the spectacular cannonade began. This may indicate that part of the electrophonic noise phenomenon, travelling at high velocity *ahead* of the bolide, may be at frequencies outside the pitch range of human ears.

Meteorite Impacts

After witnessing the spectacular visible and acoustic phenomenon of a fireball, it is often a surprise to note the insignificant effects caused by the meteorite's impact on the ground. The impact effect varies with the geocentric velocity of the meteorite, its mechanical strength and the nature of the ground at the point of impact. Most of the smaller meteorites which have their cosmic velocity retarded in the troposphere fall to the ground under gravitational attraction at about 0.1 to 0.2 kilometres per second. On hard or rocky ground a small meteorite will be found lying on the surface perhaps cracked or broken by the fall with little sign of impact on the underlying surface.

If fragments fall over grassland, the larger pieces may bury themselves a few inches below the surface, and if the ground is soft, it may be difficult for the search party to spot them. In very soft ground a fragment may penetrate more than a foot below the surface, and often the sides of the hole cave in, covering the fragment completely. However, no hard and fast rules can be applied, for different fragments from the same parent meteorite – which broke up in the troposphere – produce different impact phenomena. One small fragment of the Barwell (stony) Meteorite 1965, penetrated the asbestos roof of a local factory, and then crashed through a one-inch timber floor before its velocity was finally checked. Another fragment landing immediately outside the factory barely formed an impression on the grass surface, while a third larger fragment penetrated nearby sandy ground to a depth of 44 centimetres (17 inches).

On 1st January 1869, a fall of stones occurred at Hessle in Sweden, and many of them were recovered from the surface of a frozen lake—none of the fragments had broken the ice. With larger fragments the resulting impact may be more dramatic. The Sikhote-Alin meteorite (see page 197) which fell in Siberia on 12th February 1947, created a pit 6 metres deep with a diameter of 26.5 metres. But even in this fall a few of the craters formed by fragments weighing many tons are comparatively small. A 60-ton meteorite from the Hoba farm near Grootfontein, South-West Africa, the heaviest known meteorite, embedded itself in friable limestone to a depth of 1.5 metres. It is not known at the present time what size object created the Arizona Crater (Plate 41) or the Wolf Creek Crater (Plate 40). In the instance of the Arizona Crater, the meteorite concerned *may* be lying buried at considerable depth* as an object of this order would suffer little retardation during its brief passage through the atmosphere.

If a meteorite fragment is picked up immediately after falling, in the case of iron meteorites it may be slightly warm to touch, and in the case of the stony meteorites it will show very little sign of heat. It is a popular misconception that meteorites, owing to the great heat they develop in flight, are 'red-hot' on arrival. Before they enter the Earth's atmosphere, the temperature will be about 4°C (or 277°K). Although the outer surface rapidly heats up, the inside temperature changes very little. Their passage through the atmosphere lasts only a few seconds so that the build-up of outside heat has no time to conduct inwards. The insulation properties of meteorites are such that immediately beneath the surface the structure is unaffected by the high external temperature experienced during its descent.

There are a number of folk-lore tales about freshly fallen meteorites causing fires owing to their great heat, but there is no instance on record where this has actually occurred; and

* Although very unlikely (see also page 189).

neither grass nor straw lying beneath a fragment has ever shown signs of scorching. In some instances freshly fallen meteorites have been found covered in thin ice coatings which have formed in the atmosphere subsequent to their braking in the retardation zone. Nevertheless, when a large meteorite falls, such as the Siberian 'meteorite' fall in 1908, the body will lose little of its cosmic velocity, and in front will be trapped a compressed cloud of incandescent gas which may cause a violent explosion when the meteorite impacts with the ground. Indeed burnt vegetation and trees round the immediate area of the Siberian fall indicate too well the dramatic consequences of such events.

Meteorite Craters

If a meteorite on impact is large enough and has sufficient cosmic velocity, a meteorite crater will result from the fall. The average meteorite on landing creates little soil disturbance and will produce only an unspectacular smallish pit* or a depression if the soil is sufficiently yielding. However, the resulting impact from a large meteorite may produce a sizable crater.

Meteorite craters can be divided into two principal types: explosive craters and impact craters. Impact craters are the result of a meteorite falling at a velocity of a few hundred metres per second. Such a crater will seldom exceed 100 metres, and a characteristic feature is a banked rim which is formed by piled-up soil fragments. Explosive craters are the result of a meteorite falling at a velocity of 3–4 kilometres per second. At such high velocity at the moment of impact a shock wave spreads out from the crater and breaks up the meteoric body as well as the surrounding soil. Explosive craters range from 100 metres upwards.

It is believed that all large meteor craters at present recognized have been caused by iron meteorites. The famous Arizona Crater (see page 188) represents the explosive type crater while the Sikhote-Alin craters (see page 197) the impact type. Other large craters may have been formed due to influence of

* A meteor crater less then 10 metres in diameter is called a pit. However, the term crater is often applied loosely to all holes in the ground caused by meteorites.

both mechanisms.

Frequency of Meteorite Falls

Various attempts have been made to assess the number of meteorites reaching the surface of the Earth. As over 70 per cent of the Earth's surface is water, any that fall over the seas and oceans will be lost, and since vast tracks of the oceans have little shipping, the majority of such falls will occur without notice. An important question is whether meteorite falls tend to favour a particular region of the Earth. An examination of the recorded falls might lead one to believe that this is indeed the case. Of the 1,400 or so catalogued falls which occurred before 1940, America accounted for about 660, Europe 340, Asia 230, Australia 91 while Africa had 70. But closer scrutiny of the make-up of the figures show that these recoveries have no bearing whatsoever on meteorites favouring a particular locality.* In the example of Africa the catalogue shows that no less than 34 of them were found or seen to fall in the southern part of the continent. In Asia, out of the 230 specimens, 111 came from the densely populated Indian subcontinent. This apparent lop-sided distribution is directly influenced by the density and cultural achievement of the population. Before 1790 there were only about 20 falls recorded whereas in the following 150 years a further 1,400 meteorites were found! This is not because the frequency of meteorite falls has increased suddenly, but a reflection in the general interest and acceptance of the idea that bodies indeed do reach the surface of the Earth from cosmic space.

Taking into account the factor of unequal distribution of sightings and recoveries, a figure has been arrived at which indicates that meteorites fall at the rate of one fall per million square kilometres per year. This corresponds to a total in-fall of approximately 500 meteorites per year for the entire globe. Relating this to the areas occupied by the seas and oceans, 350 meteorites will disappear without trace. The majority of the remainder will fall unnoticed; perhaps only a score will come to the notice of authority, and out of these only half a dozen

* In South America and South Africa there are some remarkable concentrations of iron meteorites and craters which appear to have originated from large 'single event' falls.

will become available for laboratory study.

Although we can see that geographical location has no influence on the frequency of meteorite falls, there are marked seasonal and diurnal influences which introduce real effects. An analysis of the period 1800 to 1960 reveals that meteorite falls reach a peak during the months May–June while March is a period of consistently few falls. At present there are no obvious reasons why the early northern summer months should experience such peaks. It has been suggested that most observations of meteorites occur in the northern hemisphere, and therefore the period of fine weather in summer should favour their recovery, but this argument is inconclusive. Another suggestion is that the peak months are due to the Earth encountering a swarm of meteorites at this particular point in its orbit. It is of interest that this peak (May–June) is coincidental with *minimum* displays of meteoroids. Although this provides no evidence for the seasonal variations in meteorite falls, it is significant evidence when considering a possible genetic relationship between meteorites and meteoroids.

The day of the month in which a meteorite falls is not influenced by any factor. But the hour of day in which it falls shows decided preference for the afternoon hours. This can easily be explained by the motion of the Earth round the Sun, and it provides evidence about the direction of a meteorite before it entered the atmosphere. All meteorites which fall from noon to midnight have the same direction of motion as the Earth, while those falling from midnight to noon either collide with the Earth head-on or are overtaken by it (Figure 10). If the meteorite and the Earth are moving in the same direction, the entry velocity is the difference between the Earth's orbital velocity (29.77 kilometres per second) and the meteorite's orbital velocity. If the meteorite strikes the Earth head-on, the two velocities (Earth + meteorite) are added together. Because of the high velocity incurred by a head-on collision, the bolide will ablate more quickly in the atmosphere and may completely erode to leave dust fragments which remain in the atmosphere for long periods. The diurnal frequency of meteorite falls again provides useful information in respect to theories about their origins (page 220).

Hazards of Meteorite Falls

In modern times there is no authenticated record of a human

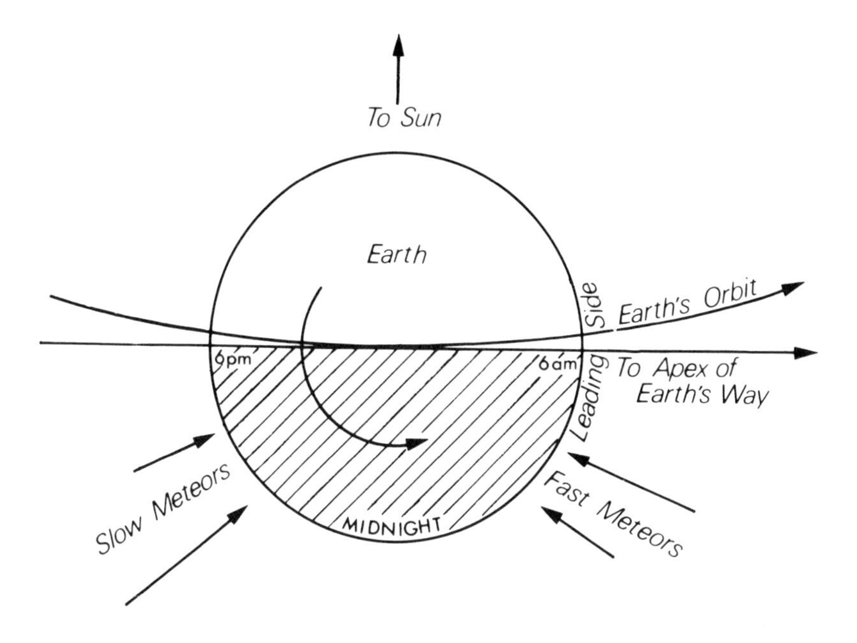

Figure 10. The diurnal effects on meteors encountering the Earth. Before midnight the observer sees only those meteors that overtake the Earth. After midnight the observer is on the leading (sweeping) hemisphere and sees the brighter, faster meteors which the Earth encounters head-on.

having been killed by the fall of a meteorite. However, there are many apocryphal stories such as the one concerning a wedding guest supposedly killed by a meteorite a few years ago in one of the Balkan countries. Perhaps the best known apocryphal fatality concerns a monk who is alleged to have been killed by a meteorite in Cremona in 1511. Strangely enough there is another similar story which relates the death of a monk in Milan about 1650, but none of these stories can be authenticated.

There seems no doubt, however, that animals have been

killed by falling meteorites. A well known case occurred in 1911 when a meteorite fell in Egypt and killed a dog; another meteorite fell in Ohio in 1860 and killed a calf. But there are numerous records of injury to human beings and damage to property. In 1954, a housewife in Alabama was struck on the arm by a meteorite which crashed through the roof and ricocheted off a radio. When a fragment of the Barwell meteorite bounced off the road, it broke a house window and landed neatly in a sitting room flower pot so that the piece was not found until several weeks later! Another piece penetrated a factory roof (see above), and another dented the body of a car.

Nevertheless, in spite of these cited instances, the chances of being struck by a meteorite are very remote. Calculations of such probabilities are necessarily speculative, but one often quoted figure is that calculated for a resident of the United States which shows that the chance of being struck by a meteorite fragment (in the USA) will occur *once* in about 9,300 years. Another probability estimate found that only one of 66 meteorites will land in a densely populated area. Perhaps this last estimate is not quite so reassuring as the first one, and on reflection it is a little surprising to what a minor extent a meteorite fall has been a potential human hazard in view of the high density of human life in present-day urban communities.

Meteorite Ownership

The ownership of a particular meteorite has on occasions involved prolonged litigation in the courts. A case heard many years ago in the United States led to the judicial decision: "A meteorite or aerolite, though not buried in the Earth, is, nevertheless, real estate, belonging to the owner of the land (on which it fell), and not personal property, in absence of proof of severance."

Another ownership dispute occurred in France in 1879. A peasant crossing a field observed the fall of a meteorite which struck the earth and buried itself only a few yards away. Although scared out of his wits, he dug up a stony meteorite of considerable size which had lodged about 60 centimetres below the surface. Guessing that it may have some value, he consulted the schoolmaster of his commune who suggested he tried the local museum—where to his joy and astonishment he received the wonderful sum of 250 francs in hard cash.

But his joy was short-lived; news of the peasant's good fortune at finding cosmic treasure reached the owner of the field who resided in Paris, and he brought an action. He claimed either restitution of the aerolite, which fell on *his* land, or 10,000 francs damages, which he judged to be the value of it. An eminent Paris advocate was retained for the peasant, but the proprietor of the land more wisely chose a local man. As they knew of no precedents, the poor judges decided to proceed by analogy, and the case was assimilated to that of alluvial soil brought by an inundation. The local advocate representing the landlord then played his trump card, and quoting the *Institutes* of Justinian, chapter "De Thesauris", Book IX: "Let no one steal the wealth buried in another man's field", swayed the tribunal to find for his client, and the unfortunate peasant was required to give up his ill-gotten 250 francs and pay all legal costs.

But if the French tribunal had truly wished to follow Roman Law and Justice, then they should have followed the example of Roman Law XXX which provides that any treasure should belong half to the discoverer and half to the owner of the soil upon which it was found. Unfortunately, however, there has always been a difference of opinion what defines a treasure. To the Romans it was "of anything valuable".

In Great Britain the Crown has recently sought, through the introduction of a Parliamentary Bill, to claim ownership as Crown property of all meteorites. At the time of writing (1972) the Bill has not yet been passed. The introduction of this Bill – with the backing of the British Museum Trustees – is a direct consequence of the problems involved in the recovery of the fragments of the Barwell (1965) meteorite. Some important fragments certainly passed directly into the hands of the professional dealers. However, perhaps this was not surprising in view of the small monetary reward (10s per ounce) instituted by the British Museum. The largest fragment that found its way to the British Museum Collection netted its finder a reward of £130. But some locals now realizing that meteorites *had* value, held on to their finds, and when professional dealers announced that they were prepared to substantially improve on the Museum's rate, the locals took advantage of this and sad to relate science was denied.

CHAPTER XIV

Meteorite Recognition, Classification and Ages

During the course of a lifetime the average casual observer will catch many a glance of brilliant fireballs as they flash across the night sky. On much rarer occasions the observer may witness a meteorite fall with the accompaniment of all the dramatic noise phenomena.

But when does the observer know for certain that a particular fireball has produced fragments which have landed in the neighbourhood? For a meteorite to fall nearby, the fireball should appear to extinguish itself high in the sky, well above the immediate horizon. When a fireball appears to glow until it reaches the Earth, or disappears low above the tree tops, one can almost certainly say that the fall has occurred at least 100 miles distant!

The fall of a meteorite in a particular locality occurs so rarely and so unexpectedly that interested professional meteoriticists will seldom, if ever, observe a single one during their lifetimes. Meteorite research, particularly that part devoted to fall phenomena, depends much on the lay public's assistance so that it is always worth-while for the interested observer to be aware of the kind of information which is scientifically useful if he is close at hand to witness a fall event.

Owing to the suddenness of a meteorite fall, the key eye-witnesses are usually taken by complete surprise, and much valuable information may be lost through inattention to detail. Since meteorites may fall at any time of day or night, the observer may at short notice be required to organize a local search or collate other eye-witness reports for the attention of national authority. The writer remembers an occasion a few years ago lecturing to an audience (in Britain) about the vast amount of meteorite research accomplished by amateur workers in the United States and concluding with a remark that there was little opportunity or scope for that kind of activity in Great Britain with its much smaller land area. Not long afterwards he found himself with a party of enthusiasts chasing around in the fields of rural England at Barwell (Leicestershire) after a meteorite had fallen there at the highly inconvenient time of 4.00 pm on Christmas Eve 1965.

The important information worth noting about the occurrence of a fall is listed below, but a piece of advice which cannot be stressed too frequently is to commit whatever information one collects to a notebook or tape recorder at the earliest opportunity. Memories are unreliable, and when an eyewitness' tale is related many times, it tends to change slightly at each telling! For interviews with key eye-witnesses the portable tape recorder is ideal, and spontaneous remarks so recorded often turn out to be much more accurate than formalized thought-out statements later committed to paper.

*Meteorite Event Questionnaire**

1) What time (or times) did the fireball(s) appear? – to be recorded to the nearest minute.
2) For what duration of time did the light remain visible? A reliable method is to count the seconds aloud (o–n–e – t–w–o, etc).
3) How bright was it: brighter than the Moon, or brighter than the Sun?
4) Did the fireball change colour during its flight path?
5) Did it explode or appear to fragmentate? If so how many pieces?
6) How long did the train (trail) remain visible?
7) At what position in the sky did it first appear (at night, position in relation to stars, by day, approximate angle of altitude and direction). The position sometimes may be related to a geographical feature such as a hill, a tall building, church steeple or chimney.

* In Great Britain the British Astronomical Association invites such information as is available to be passed on to them as soon as possible.

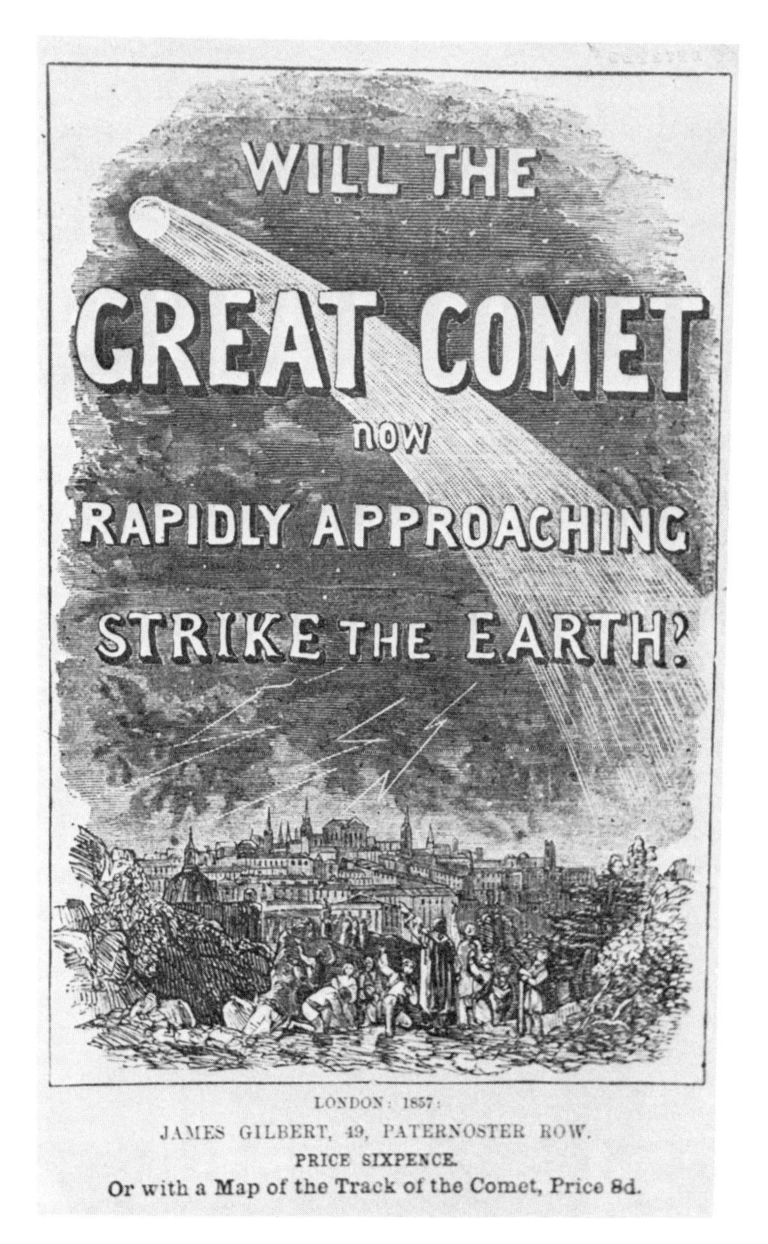

32. A sixpenny street-corner pamphlet of 1857 announcing the forthcoming appearance of a comet.

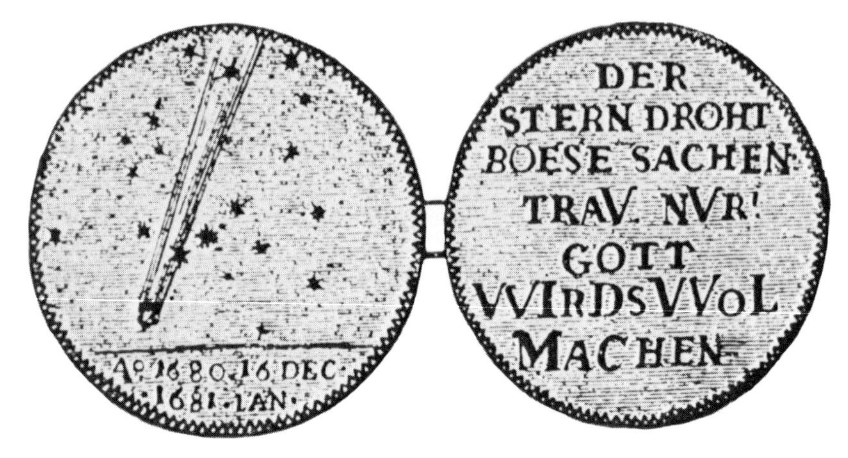

33. A German medallion struck to commemorate the 1680 Comet: "The star threatens evil: only trust! God will make things turn to good."

34. An old German woodcut print showing the famous egg laid in Rome after the appearance of the 1680 comet.

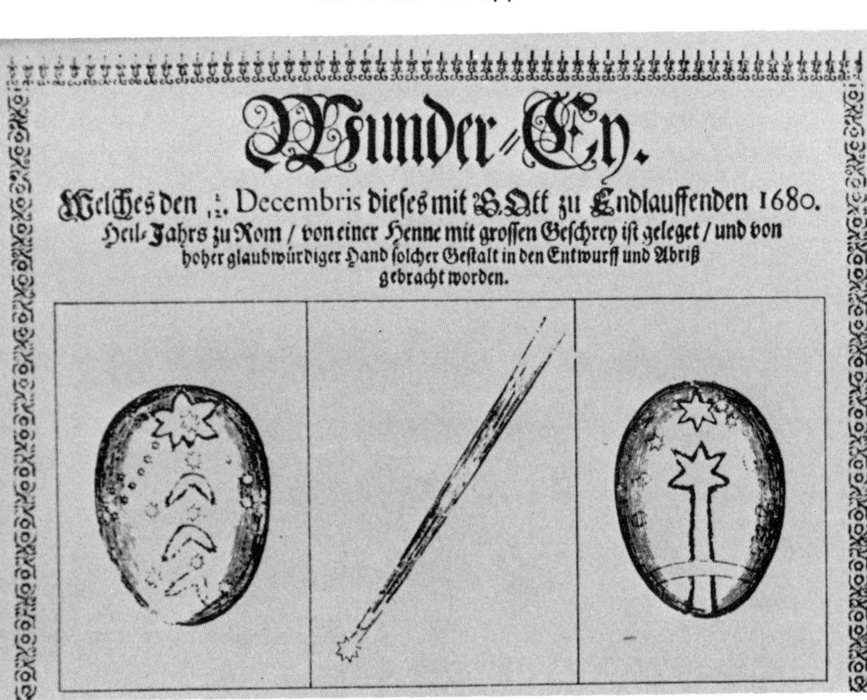

Wunder-Ey.

Welches den 2/12. Decembris dieses mit GOtt zu Endlauffenden 1680. Heil-Jahrs zu Rom / von einer Henne mit grossen Geschrey ist geleget / und von hoher glaubwürdiger Hand solcher Gestalt in den Entwurff und Abriß gebracht worden.

35. E. F. Chladni, the father of the science of meteoritics.

36. The great fireball of 24th March 1933. Later 4 kilogrammes of meteoritic fragments were recovered in New Mexico.

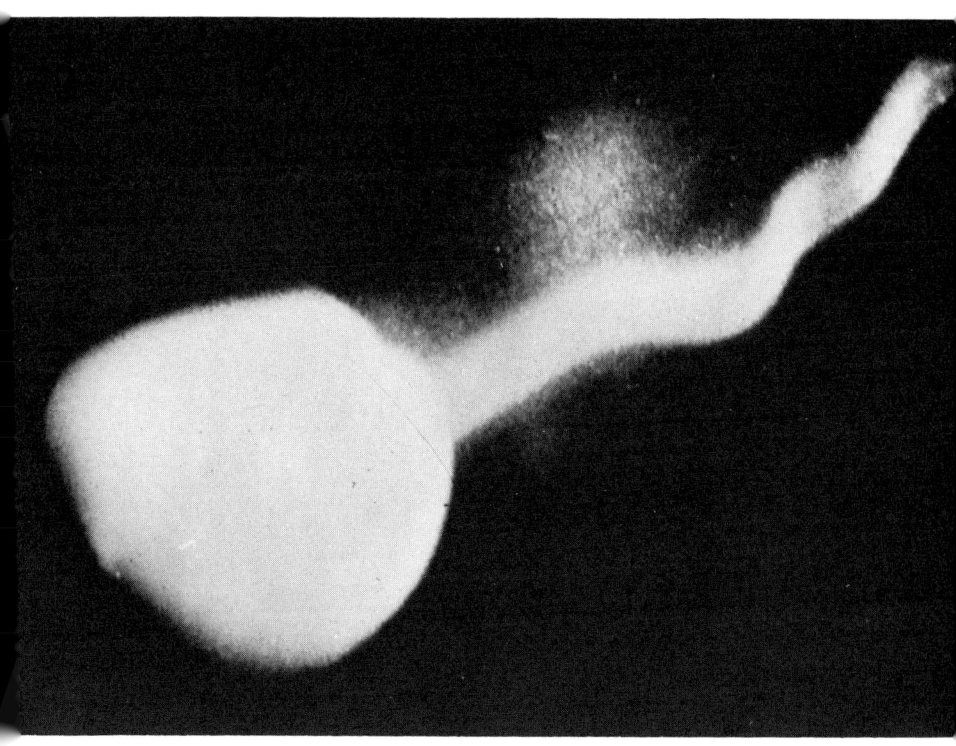

37–38. The Barwell Meteorite. *(left)* A small impact crater, and *(right)* holes in factory roof and floor decking produced by fragments of the meteorite.

39. Reassembled fragments of the Barwell meteorite: note the regmaglyphs which have the appearance of thumbmarks. The lighter-coloured interior can be seen where the thin black fusion crust is broken.

40. The Wolf Creek meteorite crater.

41. The Arizona meteorite crater.

42. Widmanstätten pattern in a meteorite. Terrestrial nickel-iron does not show this characteristic etching; in meteorites it is probably due to slow cooling in space.

43. Genuine and pseudo meteorites: (left) blast furnace slag; *(top right)* fragment of the Barwell meteorite; (bottom right) fused terrestrial iron.

8) At what position did it disappear? (apply the same rules as (7) above).
9) What sounds were heard and what kinds of sound? Cannonades? Thunder? Express train? Pistol cracks? Rustling, 'singing' or buzzing? Or choose another description if none of the above accurately describes it.
10) If sounds were heard, when were they heard: Before? At the time of the light phenomenon? or later? Estimate time in seconds.

If a meteorite, or suspect meteorite, is subsequently located on the ground after a fall, the additional particulars should be noted.

1) How long was it between the observation of the flight path and the discovery of the fragments?
2) How many pieces were found?
3) What was the penetration depth (if any)? Give details of the soil conditions.
4) Describe the physical appearance – fusion crust, inclusions, colour, etc (see below). Was it warm to touch?

The above points are only basic questions that the observer or finder should be sure of noting. Detailed professional research, followed up at a later date, can fill in the details, but this subsequent work may be of less value if the basic topical information is inaccurate or misleading.

The majority of fireball observations will never lead to immediate recovery of the bolide fragment(s) which caused it. But the fragments have often been located after deductions by experienced workers based on accurate information gleaned from amateur observers.

Fusion Crusts

The most important characteristic feature of all meteorite specimens* is a thin covering shell known as the fusion crust. It is never more than a few millimetres in thickness and will only reach greater depth where the melted material has been forced into pressure ridges or thickenings during its passage through the atmosphere. Meteorites which do not fragmentate

* Except those fragments from the inside of a highly fragmented meteorite.

either in the air or on impact with the ground, usually show a fusion crust over their entire surface area. With iron meteorites the crust is generally much thinner, and if an iron has lain on the ground for any length of time, no crust will be visible owing to the chemical action of the soil, and the atmosphere will have created in its place a layer of rust which scales off at the slightest touch.

The usual colour of the fusion crust, both on irons and stones, is black, but sometimes freshly fallen irons are found with a black crust that in certain lights suggests a slightly bluish tint, although this characteristic disappears within a few days. Occasionally the crust of a stony meteorite will appear brownish or even reddish. In rare instances a stony crust may be semi-transparent.

In addition to the fusion crust, the intact surface of meteorites show curious thumb-like impressions called regmaglyphs (sometimes called piezoglyphs) which have the similar appearances on all the principal meteorite types irrespective of their mineralogical make-up. They are formed during the flight of the meteorite through the atmosphere and are found on sides or rear ends of the body in respect to its orientation during its air passage. They do not form on the front surface because of air pressure. The side (or lateral) regmaglyphs have a different shape to the rear ones. The lateral ones are elongated (almond-shaped) while the rear ones have distinctive rounded shapes. The size of the regmaglyph provides a direct measure of the size of the parent meteorite. Generally the diameter of the regmaglyph represents one-tenth the approximate diameter of the body which produced the atmospheric fireball.

Meteorites may tumble or slowly rotate during their passage earthwards; these examples will not show a difference between lateral and rear surfaces, and regmaglyphs may be absent entirely. Some specimens are found with deep pits which almost suggest that the meteorite surface has been artificially drilled out. The cause of these deeper holes is not fully understood, but it may be due to homogeneous mineral inclusions that have melted out completely during flight; a mineral such as troilite might certainly cause this.

Meteorite Recognition

By far the greatest number of meteorite fragments are located accidently either lying on the Earth's surface or just immediately below it. Farmers often plough them up, and building workers make chance finds while excavating foundations for drains or new buildings. Unfortunately, however, the greatest number of objects which are brought to light and called 'meteorites' by the lay public are pseudo-meteorites.* Lumps of slag and waste industrial products are frequently mistaken, and natural occurring minerals such as nodules of iron pyrites or odd-shaped chunks of native metals form a large proportion of amateur finds (see Plate 43). Paradoxically it is the genuine stony meteorite which lies unrecognized, for if not quickly collected they soon oxidize and then cannot be identified as cosmic artefacts by superficial appearance alone. One sizable chunk of the Barwell chondrite lay ignored in the street for many days after the fall, since everyone thought it was only a lump of freshly broken concrete which had fallen from a passing truck!

How then does one recognize a genuine meteorite on the ground? This, of course, is assuming we are not thinking in terms of an object lying inside a small impact crater or pit which by itself would be sufficient to attract attention for closer inspection. In point of fact there is no *single* recognition criterion which is valid for all meteorites, since they assume a wide variety of forms both in physical appearance and in chemical make-up. Nevertheless, by process of elimination one can quickly narrow the field.

Any stone covered with a thin, dull black fusion crust (Plate 39) on any part of it is worthy of further examination. Many crusts will show definite flow lines where molten material had resolidified during the final stage of descent. Except for meteorites with fresh-looking fusion crusts, the outside appearance is not a good guide except to the very experienced. But a suspected meteorite must *not* be broken into smaller pieces. The best single test is to grind a small area of the specimen on an abrasive wheel. If metallic iron is present, it will be readily visible in the stony mass, and its presence is very good evi-

* Perhaps one specimen in several hundred submitted for expert opinion is a genuine meteorite.

dence indeed. A stony surface may contain metallic specks which to the eye will glint like silver if held towards the Sun or a strong light. Iron meteorites are all magnetic, and if a suspicious stone* shows this indication, it is a useful guide but not conclusive. On exposed surfaces the free iron will quickly rust. A genuine meteorite *never* shows gas vesticles such as those seen in specimens of industrial slag wastes (Plate 43) which are frequent sources of pseudo-meteorite finds. Neither do meteorites have characteristics similar to volcanic laves with open structures. Iron meteorites are about three times heavier bulk for bulk than terrestrial rocks, while stony meteorites are only slightly heavier.

Although meteorites may assume any shape, a conical-nosed artefact covered with a black dull fusion crust and flow lines is most certainly a meteorite. The nickel-irons are probably the easiest meteorites to confirm, for they contain a percentage of nickel ranging from 5 to 20 per cent, while terrestrial iron contains nickel in larger or smaller quantities. Although with elementary chemical knowledge it is easy for the interested field-worker to detect the presence of nickel with a few inexpensive chemicals, the method is not included here since it is wiser to hand over suspect objects to suitably equipped chemical and physical laboratories where they can be examined professionally. Indeed, if any suspicious, non-terrestrial looking stone is found which, after applying the above basic criteria, still creates doubt, it should be handed over to the local museum for expert evaluation.

Meteorite Chemistry and Minerals

Until man landed on the Moon and space probes reached Mars, meteorites were the only source of cosmic material which man could study in a terrestrial laboratory. Meteorites, like terrestrial rocks, are aggregates of minerals, consisting of elements in varying percentages. The study of material from outside our own environment is an exciting occupation. An intriguing three-part question which the study of meteorites raises is: Do meteorites contain elements unknown on Earth, minerals unknown on Earth, and living organisms?

* On account of the minute grains of nickel-iron that are scattered randomly throughout the interior of stony meteorites, they are commonly more susceptible to a magnet than a terrestrial stone of the same size.

The answer to the first part of the question cannot be a categorical no. Of the 92 elements in the periodic table – not including the transuranium elements – the only undetected members in meteorites are technetium, promethium, astatine and francium. But one of the isotopes found in relative abundance in stony meteorites is xenon-136. This is the product of radioactive decay, but no known chemical elements could produce it in the proportion in which it is found in meteorites. The parent element is considered to have been highly volatile. However, the only elements like this are *theoretical ones* which occupy a position between 112 and 119 on the periodic table. A model of such an element's atomic structure has been calculated by computer, but attempts to synthesize it in a terrestrial environment have been unsuccessful.*

The answer to the second part of the question is yes. It is an accomplished fact that minerals unknown on Earth have been discovered in meteorites, but so far in very small amounts, and such minerals can actually be produced artificially in terrestrial laboratories.

The answer to the last part of the question is probably no. This must be qualified by the statement that there is considerable disagreement about it. Certain meteorites contain elements that suggest organic substances, but the detail discussion of this evidence is provided for elsewhere (see page 222).

The most abundant chemical elements found in meteorites are iron, nickel, sulphur, magnesium, silicon, aluminium, calcium and oxygen. Iron appears to be the most important component of all meteorites, for even the stony meteorites average about 15 per cent iron by weight. Nickel is next in importance, and this is encountered in combination with iron to form nickel-iron. The stony meteorites are also rich in silicon, containing about 20 per cent by weight, and in addition magnesium is important, representing 12 per cent by weight. The stony-irons, as one would expect of an intermediate class, contain the same dominant minerals but in different proportions.

Of the more exotic minerals, diamonds were first discovered in a stony meteorite which fell in the Soviet Union at Novyi

* The heaviest element known, element 104, was synthesized in the United States in 1969.

Urei (Nizhegorod) on 14th September 1886. Since then non-commercial diamonds have also been found in other stones and in certain iron meteorites. The presence of diamonds in meteorites appears to be well known among the lay public and has the result of causing meteorite specimens to be concealed from authority in the mistaken assumption that these diamonds may be of considerable value. Gold, silver and platinum occur frequently in many specimens but in very small amounts.

Some meteorite minerals are difficult to preserve under terrestrial conditions. For example lawrencite, abundant in many specimens, is unstable. When exposed to air, it will deliquesce into a green liquid which then changes colour into blood red. All specimens containing lawrencite need special preservative treatment in super-dry environments.

Meteorites contain a proportion of gases, the principal ones being hydrogen, carbon monoxide, carbon dioxide, nitrogen and methane. The most important gases, however, are the primordial noble gases such as helium and argon which are used to determine the ages of meteorites (see page 186).

Meteorite Classification

Three principal kinds of meteorite are now recognized: irons (*siderites*), stony-irons (*siderolites*) and stones (*aerolites*). Each of these principal three divisions can be subdivided in many subdivisions, and about 80 different classifications form the family as a whole, although authorities are not in universal agreement in the detailed 'family' tree arrangement of the various classifications.

The iron meteorites are those whose metallic constituents predominate over all others. They form three main subdivisions: *octahedrites, hexahedrites,* and *ataxides*. Some include *lithosiderites* in this group, but they really belong to the stony-irons.

The *octahedrites* are the commonest type of iron meteorite and are so named because they show the now classical Widmanstätten structure (or pattern) named after the Austrian Widmanstätten, who in 1808 first published an account of it. Actually, if justice be done, it should be called the Thompson pattern after the Englishman who first drew attention to it in 1804, after experimenting with fragments of the

Pallas iron. The structure is seen when certain meteorites are heated in a flame and then etched with acid. Under a microscope it shows a series of lamellae of two different constituents: a mineral called kamacite (which is more vulnerable to acid attack) and taenite (which is less vulnerable). The etching effect is the result of the differences in acid resistance. The name *octahedrite* is derived by the fact that the plates of the kamacite which are surrounded by their coating of taenite are arranged parallel to four pairs of faces of the octahedron. The Widmanstätten pattern is intriguing in that it does not occur in any natural earth metals. Its presence in cosmic nickel-irons is most probably the result of very slow crystallization in a space environment.

The subdivision called *ataxides* – which means devoid of order – is richer in nickel content than the *octahedrites* and this results in the absence of the Widmanstätten structure.

The *hexahedrites* are so called because they consist of cubic crystals of kamacite. When their surfaces are polished and acid-etched, a delicate series of parallel bands known as Neuman lines make their appearance. The presence of these lines is likely to be the direct result of the parent meteorite suffering mechanical deformation in environments subject to low space temperatures – probably during an explosion or collision impact event.

The transitory stony-irons (siderolites) form a minority group. These can be divided into four subdivisions: *pallasites*,* *mesosiderites*, *siderophyres* and *lodranites*. Only one example of the last group is known at the present time.

The stones (*aerolites*) form the largest representative group and are divided into *achondrites* and *chondrites*. But even the stones often carry appreciable amounts of nickel particles.

The *chondrites* are the most numerous class among the stones. The name is derived from the Greek *chondros* (grain), and it is the 'grains' or chondrules which are the most distinctive physical characteristic of their structure. These chondrules vary in size from the microscopic to the size of a pea, or sometimes larger. Only the smallest sizes have perfect spherical forms, and the group as a whole may vary in size and colour (Plate 47).

* Named after the explorer/naturalist Pallas (see page 154).

Usually the chondrules are embedded within a granular mass and can be separated from it without breaking. Practically all minerals found in stony meteorites can form chondrules. The theory of chondrule formation in chondrite meteorites has not been universally agreed upon (see page 228).

The *achondrites* lack the characteristic spherical chondrules of the above. Some of the sub-types resemble terrestrial igneous rocks, and this make field recognition less easy.

Meteorite Types

The greater proportion of meteorites found on the Earth's surface are irons ranging in weight from a few grammes up to several hundred metric tons, while the stones account for only one quarter of the total. Nevertheless, among observed falls stones represent 80 per cent, but irons only 6·5 per cent. The reason for this apparently contradictory situation is that stony meteorites resemble certain terrestrial rocks, and because of their greater tendency to weather, they escape detection more easily if they are not recovered soon after falling. The blocks of nickel-iron, however, resist weathering and are generally so conspicuous that they are easily recognizable.

Tektites

Some of the most puzzling objects found in widespread locations all over the Earth's surface (Figure 11) are small glassy objects (consisting of 70–80 per cent SiO_2) which at first glance superficially resemble the terrestrial volcanic rock obsidian. On closer inspection, however, they are found to be quite different and have the appearance of small distorted globules of liquid rock which has solidified in flight.

These objects have been given the group name tektites, derived from the Greek *tektos* meaning 'molten'. No terrestrial rock found *in situ* resemble them, and neither do they bear any affinity to the local country rocks in the various locales where they are found lying on the surface.

They have been recognized for over two centuries as strange objects, and each group takes its particular name according to the geographical locality where found. The tektites found in Czechoslovakia are called *moldavites* – after the river Vltava

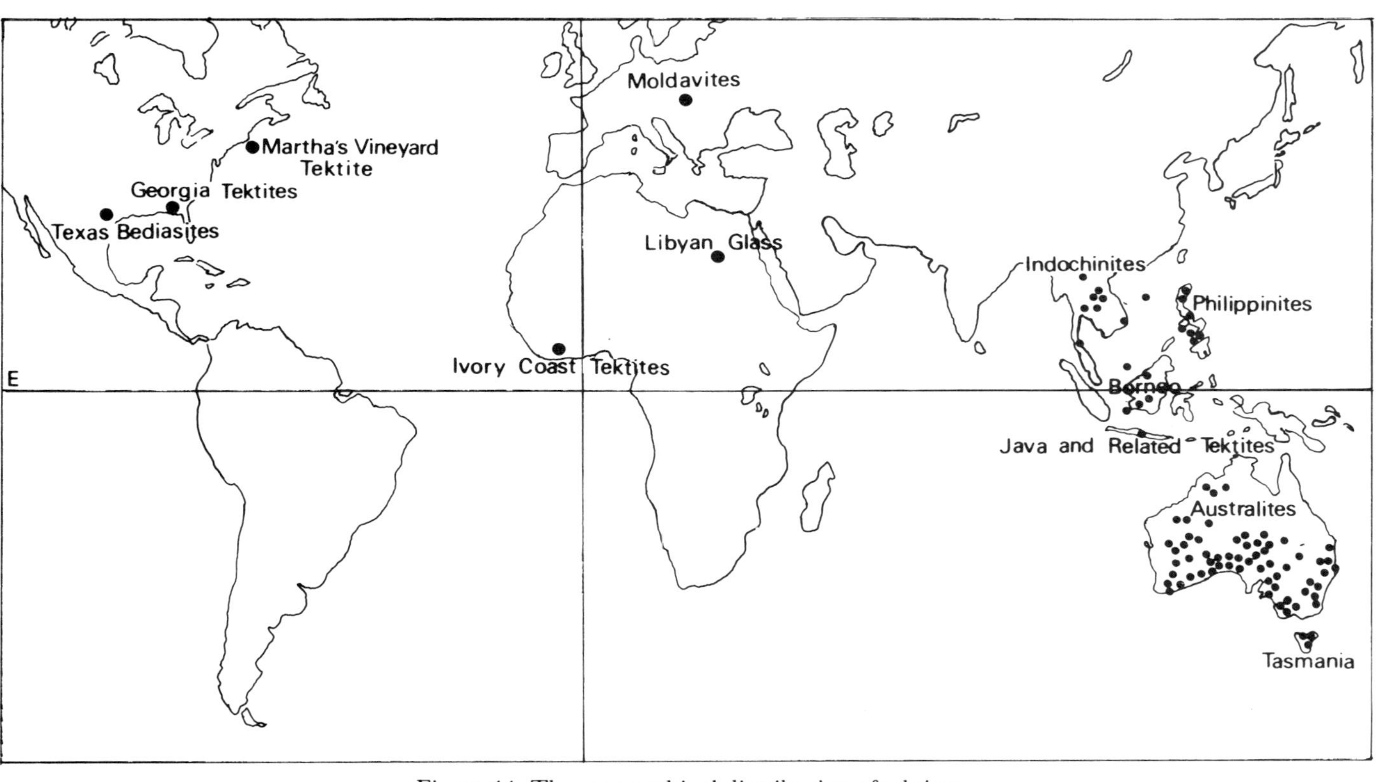

Figure 11. The geographical distribution of tektites.

(German name: Moldau) where large concentrations are found. The group known as *australites*, found in Western, Central and South Australia, have long been revered by the Aborigines as 'sky charms'. They are found in large numbers by Australian prospectors using 'dry blowing' alluvial mining techniques. The *australites* are quite distinctive owing to their unusual shapes – they appear like buttons, mushrooms and miniature hour-glasses.

The characteristic shapes of tektites can be explained by aero-dynamic modelling if they truly represent objects which were once liquid droplets and were solidified during flight. They range in colour from black through brown to a kind of bottle green, and range in size from a walnut (or smaller) to an apple.

Present-day opinion is still divided about their origins. It may be that they are terrestrial in origin and are the result of volcanic activity; or they may be the product of meteorite (or cometary) impact – either in the form of modified cosmic material or crustal material transformed (into tektites) by heat. Nevertheless, weight of opinion at present favours the meteorite (or cometary) origin. A previous idea that tektites may originate from the lunar surface has now lost favour. Marine sediments examined in the cores taken in the South Pacific Ocean contain microtektites (diameter 1 millimetre) which resemble the nearby Australian mainland tektites. These latter objects have been accurately dated using modern laboratory techniques and are approximately 700,000 years old.

Micrometeorites and Micrometeoroids

Meteorite particles only a few microns* in diameter constantly bombard the upper atmosphere of the Earth. But they are so small that they are quickly retarded while still at great altitude long before most of them reach the lower burning (or ablation) zone with the result that the largest percentage reach the surface of the Earth intact, having suffered little or no frictional heating.

* One micron $=10^{-3}$ millimetres.

The term *micrometeorite* is normally restricted to the small particles unaffected by their passage through the atmosphere. *Micrometeoroids* are particles which have become partially melted owing to some frictional heat being created during their fall. Unlike *meteorites* (see above) and *meteoroids* ('shooting stars' see page 203) these smaller cosmic particles are never visible to the human eye. Many of these are so tiny and fragile that they remain suspended in the upper atmosphere of the Earth by thermal currents for long periods before finally settling on the Earth's surface.

They have been collected in the atmosphere by high-flying aircraft using a device appropriately named the "Venus Flytrap" which works like a kind of aerial plankton net. Orbiting satellites have also detected them on specially exposed stainless steel plates, which afterwards show tiny craterlets where the tiny fast-moving particles have impacted and partially melted. On the surface of the Earth they are found in deep ocean sediments and trapped in the ice of the Arctic and Antarctic ice caps. Some micrometeoroids, when examined under high magnification, appear to have irregular, loose fluffy structures with almost a gingerbread texture, but others appear to be spherical balls of tougher material. Some of the particles trapped by the high-flying aircraft and rockets are also spherical (black-coloured) objects, but these are probably of terrestrial origin and represent waste products from industrial processes connected with iron and steel smelting.

Meteorite Ages

Terrestrial rocks can be aged by the rate of decay of their radioactive minerals, and in a similar fashion ages can be derived for meteorites. The method is to measure the radioactive nuclides in small amounts so as to determine the concentration of the parent and (decay transformation) daughter nuclides on the basis of known determinations for their half-lives.* In theory the idea is beautifully simple; in the laboratory the process is fraught with untold practical difficulties.

* Thus of a given quantity of Rubidium, half will have decayed to Strontium after an interval of 47×10^9 years.

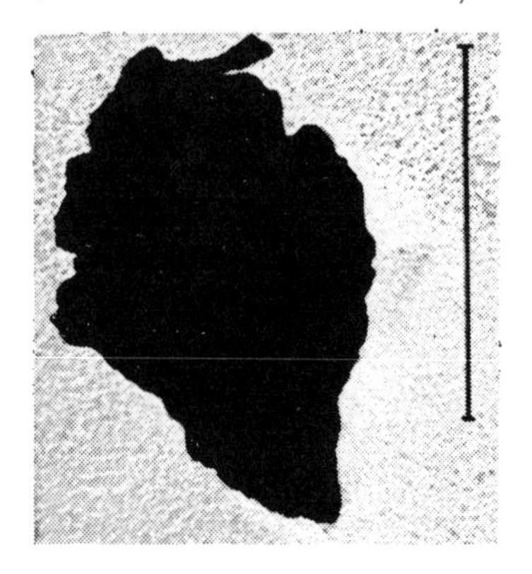

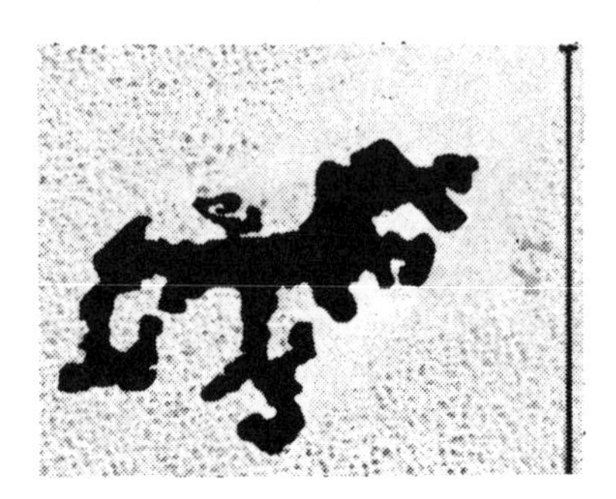

Figure 12. (*left*) A 'solid' micrometeorite; (*right*) a 'fluffy' micrometeorite. Note scale of one micron (1μ).
($1\mu = 10^{-8}$ milimetres)

At present there are five principal processes in use to arrive at radiogenic decay ages:

Parent Nuclides		*Daughter Nuclides*	*Half-life* (yrs)
Uranium	(U^{238}) to Helium	(8He)+Lead (Pb^{206})	$4{\cdot}51\times10^9$
Uranium	(U^{235}) to Helium	(7He)+Lead (Pb^{207})	$0{\cdot}71\times10^9$
Thorium	(Th^{232}) to Helium	(6He)+Lead (Pb^{208})	$13{\cdot}9\times10^9$
Rubidium	(Rb^{87}) to Strontium	(Sr^{87})	47×10^9
Potassium	(K^{40}) to Argon	(Ar^{40})	$1{\cdot}25\times10^9$

The potassium-argon method is used for stony meteorites since a loss of helium occurs in stones, while the helium method is used in only the irons, for these do not contain sufficient amounts of potassium.

Using the results of these radioactive decay mechanisms provides ages ranging from 5 aeons* to 4·5 aeons. Most meteorite ages, however, show an average of 4 aeons. The approximate age of the Earth is at present estimated at 4·6 aeons.

The ages of meteorites can also be studied by another method which is not available in the study of terrestrial rocks. Every object in space is constantly under bombardment by high energy cosmic rays. Although the Earth is subject to the

* One aeon = one thousand million years.

same bombardment, the atmosphere protects the surface rocks from primary cosmic nuclei in meteorites and knock out some of the protons and neutrons, changing the identity of the atomic structure and so leaving a record. The additional advantage of dating by cosmic rays is that the rays cannot penetrate below a depth of about one metre of rocky material. Thus by comparing the radioactive decay age with the cosmic ray age, we can arrive at a date when the particular chunk of meteorite was shattered from its parent or sub-parent body and so became vulnerable to cosmic ray bombardment. It can therefore be seen why derived cosmic ray ages are much shorter than ages derived by radioactive methods. But average cosmic ray ages for stone (0·2 to 500 million years) are different to average cosmic ray ages for irons (100 to 1,500 million years). It is believed that this difference is due to the stony meteorites being structurally more vulnerable, and they break up more readily in space than the iron meteorites.

It is also possible to age a meteorite in terms of the time elapsed since it fell to the Earth, and this is referred to as its *terrestrial age.* After it reaches the Earth, a meteorite becomes shielded from further cosmic ray bombardment, and its radioactivity begins to decay. The amount of decay then provides its age. By such a method it has been shown that some of the well-known irons fell to Earth many thousands of years ago, and one example is perhaps 500,000 years old.

CHAPTER XV

Famous Meteorites and Fossil Craters

One of the most famous landmarks of the American West is the direct result of a huge meteorite which fell in prehistoric times at Canyon Diablo in the northern Arizona Desert. The impact gouged out a huge crater 1,260 metres (4,200 feet) wide and 170 metres (570 feet) deep, which has a prominent rim rising 50 metres (165 feet) above the surrounding country. It is estimated that 400 million metric tons of rock were displaced by the fall.

It was first noticed by white men in 1871, although it had been well known to local Indians for many centuries. It was visited by geologists in the early 1890s, but they were spectical about suggestions that it was meteoric in origin. A mining engineer named Daniel Moreau Barringer had no such doubts; the finding of numerous nickel-iron fragments which were widely scattered over the surrounding area provided something more than circumstantial evidence. In 1902, Barringer formed a partnership with B.C. Tilghman to undertake a survey and begin drilling operations to find the remnant meteorite. A calculation based on the size of the crater led to a speculation that a valuable nickel-iron body of some 10,000 metric tons lay buried beneath the floor of the crater. But the first drilling failed to locate any sizable body.

Some early sample analyses of scattered iron fragments showed substantial percentages of gold, platinum, iridium and microscopic diamonds which at the time caused great excitement, but later more accurate chemical analyses showed only moderate traces of these minerals.

In 1908, Barringer, after rethinking the problem, decided to carry out simulated experiments using bullets as miniature meteorites. These experiments led him to the conclusion that the meteorite had impacted at an angle of 45°, and therefore the remnant mass would not be found immediately below the crater centre but displaced towards the rim of it. By 1922, Barringer was confident that he knew the exact location of the nickel-iron and floated a mining stock company in order to raise capital for another prospecting venture. New drill holes were started, and shafts excavated in the supposed direction. However, at a depth of 414 metres (1,380 feet) the drill suddenly struck hard, impenetrable material, and further attempts to deepen the hole met with failure. The exploratory shafts, sunk directly above the supposed location, met with ground water and dangerous quicksands, and pumping could not keep pace with the inflow. After half a million dollars had been swallowed up in exploration, the prospect was abandoned as unworkable.

In subsequent surface exploratory work, thousands of nickel-iron fragments have been found in the area ranging in size from a few grammes to over half a metric ton; some fragments have been found within the crater itself. There seems to be little doubt among geologists – except for one or two critics – that the crater is meteoric in origin. Nevertheless, it is very doubtful if any sizable chunk of material as envisaged by Barringer actually exists beneath the crater floor. Ideas about large cosmic impacts have changed since Barringer's time. It is now believed that the large crater is the result of an explosive/impact event, so that when the huge fireball descended, most of the solid material was vaporized, and what escaped complete destruction now lies strewn about the area in small fragments.

One of the theories advocating a terrestrial origin for the crater places its age at 200,000 years. The theory considers that the crater was formed as a recent caving-in of a dome-shaped geological fold. The presence of meteoric fragments in and about the crater is explained, somewhat glibly, by the assumption of a *second* coincidental meteorite fall to the north-east of the crater.

The meteorite certainly fell before recent historical times, and estimates range from 5,000 to 100,000 years. It is said that local Indians had an old legend which told of a fire-god descending in a flaming carriage upon the Earth, but this piece of apocryphal folk-lore must be purely circumstantial.

Formerly the crater was known as the Canyon Diablo or Coone Butte, but now it is known as the Barringer Crater in memory of one of its first champions. A small museum is established at the crater, and the crater itself is now exploited as a popular tourist attraction. Seen from the air there is a striking resemblance to craters of comparable dimensions on the Moon. Many of the leading museums of the world possess fine nickel-iron specimens from the cosmic body which produced the great Barringer Crater.

When it became generally accepted that the Barringer Crater was meteoric in origin, the search began to find other craters or depressions which may have been produced as the result of a meteorite fall. In 1921, a group of four craters was found in a remote arid region near the town of Odessa in Texas. The largest measured 162 metres (535 feet) across and 5.5 metres (18 feet) deep. The three lesser craters were not visible as surface features and were found in later excavations. The area revealed many iron fragments that showed extreme weathering and provided evidence that the craters were very old. Indeed all the alluvial infills of the craters contained remains of long extinct animals.

In 1931, a group of 13 craters was found at Henbury in Central Australia. Again these craters appeared to be the result of a meteorite shower of nickel-iron objects, all falling within an area of about 1.25 square kilometres. The region is arid semi-desert country with low rainfall. During occasional wet periods, the craters become filled to the brim and afterwards retain the rainwater to a surprising degree; long before European settlers arrived in the area they were renowned as a reliable source of water by local Aborigines.

These craters have been dated to about the same time as the Barringer Crater, and in common with this latter crater there is a remarkable Aboriginal folk-lore legend which tells that the area is to be feared. The story paraphrased into pidgin English simply says "Sun walk fire devil rock". However, too much reliance must not be placed on the idea that folk-lore legend has passed down the story from the time of the meteorite actually falling. Outback Australia is full of such Aboriginal legends which certainly have no tangible astronomical connections.

At Boxhole station, 300 kilometres (190 miles) north-east of Henbury, is another Australian meteorite crater about 175 metres (580 feet) in diameter, first identified in 1937. Nearby are the scattered remains of shale balls and other highly weathered metal fragments indicating another fall of great antiquity.

Crossing the Australian continent into Western Australia, there is a small but very interesting crater at Dalgaranga, north of the sheep and gold-mining township of Yalgoo. This is a crater 20 metres (66 feet) in diameter and about 3 metres (10 feet) deep. It was produced by a meteorite impact in weathered granite rocks located in an auriferous region, and there is no indication of an impact-explosion having occurred. This crater has only been known to geologists since the 1920s as a possible meteorite crater, and locally, for many years, it was considered to be just "a hole in the ground" remaining from an unpromising open-cut mining prospect which had been abandoned during the 1890s gold-rush. When the author visited the area in the early 1950s, metallic fragments were not difficult to find. Later visitors have concluded that the meteorite which caused the crater was predominantly a stony one with a few metallic nodules embedded within it, and this assumption certainly fits in with the author's earlier impressions.

Still in Australia, and moving to the Kimberly district in the northwest, a large crater was photographed in 1937 at a remote spot 130 kilometres (80 miles) south of Halls Creek. It measures 860 metres (2,840 feet) in diameter and is about 50 metres (165 feet) deep. There is considerable alluvium infilling in the crater which from the air suggests the impression of an ancient lunar ring feature. The crater lies in some of the most inaccessible territory in the entire continent and was not visited by a field party until 1948. All the fragments totalling about 65 kilogrammes were found to be highly oxidized, and the crater itself is undoubtedly the product of an explosive event rather than a simple impact.

In a continent as vast and as sparsely populated as Australia it is likely that some other craters lie waiting to be discovered. In recent years the number of ancient iron meteorite finds has greatly increased since mining corporations began large-scale geophysical prospecting surveys. The northwest region, in particular the Canning Basin and the areas bordering it, is still relatively unknown country as far as ground surveys go. The author, during the 1950s, heard many first-hand accounts from boundary riders and dingo shooters about objects which resembled large, weathered irons, lying in remote spots. To the best of the author's knowledge these irons remain unconfirmed by present-day travellers in the area, but it is of interest that in times past it was not uncommon to see meteoric iron specimens in the hands of the local Aborigines, for they were long recognized as very unusual natural objects.

In 1927, six large meteorite craters were identified on the Island of Oesel in Estonia (now part of the USSR). The largest crater measures 110 metres (360 feet) in diameter and was produced by a metallic body.

During the exploration of the Rub-el-Khal Desert, Arabia, in 1932, the explorer J.B. Philby discovered two craters near Wabar which were hardly more than depressions in the shifting sands. Nevertheless, they are undoubtedly meteoric in origin, for the results of an explosive event were still plainly visible. The instant head had melted the sand to produce an area of glass foam, and chunks of silica glass were interspersed with several pieces of meteoric iron. The largest crater is about 100 metres (330 feet) across and 12 metres (40 feet) deep. The smaller one is about half this diameter. Few people have ever visited this part of the lonely quarter, and this fall is at present the least studied among the large meteorite craters of the world.

The Tunguska Meteorite

During the twentieth century the two greatest meteorite falls of the period have both occurred in Siberia. The first fall known as the Tunguska meteorite is one of the most remarkable occurrences to be studied by modern science.

In the early morning hours on 30th June 1908, eye-witnesses saw a dazzling bright fireball of immense size ploughing across the sky from south-east to north-east, leaving a pall of thick dust in its wake that soon developed into a gigantic smoke pillar. At Kansk, on the Trans-Siberian Railway, 600 kilometres (370 miles) from the impact zone, the passengers of an express glimpsed the brilliant fireball as it swept across their north-eastern horizon and described it to be so bright that it outshone the Sun. Shortly after disappearing, a great thunder-clap was heard followed by several others. So loud was the noise that the engine-driver, believing part of his train to have exploded, applied his brakes and stopped.

The impact occurred shortly after 7 am local time, and the detonations were heard over a distance of 1,000 kilometres (600 miles). A gigantic aerial pressure wave swept out from the point of impact so that it was recorded as far away as England on six microbaragraphs. In England at the time, these waves apparently caused little comment; they were simply dismissed as anomalous waves from an unknown source, and not till 20 years later was their true significance realized. But at Kansk, several witnesses were knocked off their feet by the air wave, and at Vanovara trading station 65 kilometres (40 miles) south-east of Tunguska, one isolated witness was hurled in the air a distance of several metres and felt an overpowering hot air blast before losing consciousness. Throughout the sparsely populated neighbourhood, windows were broken inwards by the pressure. In the nights following the event, a remarkable atmospheric phenomenon was widely noted extending south to the Caucasus Mountains, and it had some effect as far away as the British Isles. Several million tons of meteoric dust – dispersed by the fireball during its descent and now suspended high in the atmosphere – created a reflecting medium to the rays of the Sun just hidden below the northern horizon. The nights were so brilliantly illuminated that newspapers could be read outdoors at midnight. During a 2½-month period following the fall, the legendary white nights of the north shone more brightly than ever before or since. But northern Europe at that time had not the slightest inkling as to what had caused this unusual phenomenon.

It may seem surprising that the first scientific expedition to investigate the circumstances of the fall did not reach Tunguska until 1927, some 19 years later. It was not until shortly after 1920 that news gradually reached the world at large that a cataclysmic event had occurred somewhere in Siberia during the summer of 1908. Attempts were made in the early twenties to piece together eye-witness reports which in Czarist times had been dismissed as nonsense. The events of the Russian Revolution had been so far-reaching that scientific investigation was practically non-existent in the earliest years of the Soviet Republic. However, the Tunguska fall was obscured to a large extent by the utter geographical remoteness of the permafrost region where it occurred. When L.A. Kulik began preparations for his first expedition in 1927, he was greatly assisted in pinpointing the search area by an earlier report collated in 1924 by S.V. Obruchev during a pioneer journey along the basin of the Podkamennaia Tunguska and Angara rivers, and by eye-witness narratives collated by I.M. Suslov from local Evenkian nomads who in 1908 were camped on the *taiga* some 40 kilometres (25 miles) from the impact zone.

These eye-witness reports spoke of the annihilation of a herd of reindeer grazing within 10 kilometres (6 miles) of the impact point. The nomads' own tents were lifted bodily in the air along with their sleeping occupants, and an entire surrounding forest was demolished by the force of the air blast.

Shortly after penetrating the region, Kulik's party saw for themselves that these stories were substantially true. The forest area was completely laid bare. Uprooted tree trunks pointed radially outwards from the direction travelled by the blast while all roots were orientated inwards, providing the miserable scene with a paradoxical touch of regimental orderliness (see also Plate 46).

This evidence was sufficient to convince Kulik beyond any doubt that here was the scene of some unique event never before experienced within historical memory. At the centre of the region, in an area extending some 7-10 kilometres, were scattered many crater-like holes filled with water and a layer of sphagnum mosses. Kulik identified the holes with meteorite craterlets on the assumption that the impact phenomenon was due to a swarm of small individual bodies. But the whole area formed a huge shallow depression and swamp which alternated between stretches of clear water and sphagnums.

In 1928 and 1929–30, two more expeditions under Kulik revisited the area to excavate the small crater-like holes in the expectation of recovering meteorite fragments. But none were found, and in subsequent expeditions these 'craters' were attributed to natural processes occurring above the level of permafrost.

On the second expedition, exploratory holes were drilled and some shafts sunk to a depth of 34 metres (112 feet), but not a single meteoric fragment was found. On this expedition it was noticed for the first time that some damaged trees, partly sheltered by natural depressions, remained standing like gaunt telegraph poles. Their exposed tops had been blown off as clearly as if done by a guillotine. A few trees showed signs of continued growth at the lower, sheltered level.

An aerial expedition 1938–39 obtained photographs of the entire area, but there was no trace of a crater, and apart from the flattened forest there was little indication that a giant meteorite had fallen. E.L. Krinov, who took part in many of the later expeditions, felt that in his opinion the absence of any clear-cut features was due to the nature of the surface swamp and underlying silt. It was not until 1957 that meteoritic dust was first isolated from soil samples which had been brought back by Kulik's earlier expeditions – they had lain forgotten in a drawer!

Another expedition visited the scene in 1958, and in 1961 the Soviet Academy of Sciences mounted the most ambitious one to date, sending 80 specialists to spend four months in the area. Apart from the discovery of small quantities of magnetite and silicate – considered to be dispersed matter from the explosion – there is little chemical evidence about the probable nature of the cosmic body which caused the fireball and the subsequent catastrophic explosion.

Perhaps it is not surprising, in view of the lack of definitive information, that there is much scientific (often fanciful) speculation and controversy about the entire episode surrounding the Tunguska 'meteorite' and its fall.

There can be no doubt that the fireball possessed a very high cosmic velocity, probably in excess of 60 kilometres per second, but from here on ideas are more speculative. Three main scientific theses have been put forward (disregarding the more improbable ideas connecting it with interplanetary or intergalactic space-time travellers):

1) It was a fragile meteorite that completely exploded and disintegrated on impact, or was pulverized in the atmosphere before reaching the Earth's surface.
2) It was a comet.
3) It was an interstellar object composed of anti-matter, or a similar object which triggered a nuclear explosion on impact.

Owing to the lack of 'hard' factual evidence, any one of these ideas may be used to fit the circumstances. However, adhering to the spirit and principles of Occam's Razor (see page 229), the first is the obvious choice. It is recognized that explosive meteorite impacts could easily produce all the phenomena observed both in the atmosphere and on the ground.

The second idea is an interesting one and is certainly worthy of careful consideration. A fragile monolithic body with volatile constituents, dissipating most of its bulk during its flight through the atmosphere, fits in reasonably well with the Whipple, icy-conglomerate model of a comet nucleus. But it would also fit in with the idea of the 'sandbank' comet model if the particles forming the concentrated nuclear region had sufficient concentration of bulk to penetrate the atmosphere. Neither is it necessary to think in terms of a large comet; a Sungrazer split-off (a mini or 'pygmy' comet) perhaps so small as to be invisible even when passing close to the Earth is quite adequate to meet the requirements. These attractive ideas, nevertheless, remain highly speculative, for we have seen that little is known about the nature of comets and *nothing* is known about their likely behaviour if the Earth should happen to collide with one.

The third idea is the most bizarre of the three and is far-reaching in its implications. The idea of anti-matter has permeated scientific thought since Nobel Laureate P.A.M. Dirac* predicted it on theoretical grounds in the early 1930s.

* P.A.M. Dirac formulated the idea of the coexistence of two kinds of matter, one of which is the mirror image of the other: therefore there should

Since then many theories and ideas have evolved which discuss the consequences if such matter truly exists. The idea has been used to explain otherwise inexplicable phenomena which occur at the furthest boundaries of the known Universe and also phenomena close at hand, namely, the destroyed forest lands of Soviet Siberia.

If such matter did impact with the Earth on the morning of 30th June 1908, it would leave no chemical trace, for the bizarre consequences of anti-matter is that when it meets ordinary matter, it annihilates itself, giving birth in the process to gamma rays and ordinary radiation. The notion that the Tunguska object may have been anti-matter dates from the time when Dirac's theoretical prediction was verified after the anti-proton was artificially created in particle accelerators in the early 1960s.

The story of the Tunguska object is by no means ended. Nowadays barely a month goes by without some reference to it in the scientific press.* But until further, more concrete, evidence comes to light, the most realistic evaluation of the phenomenon leans towards the acceptance that it was a meteorite which encountered the Earth. Nevertheless, no matter what the object was that struck the Earth so violently, it is a chilling thought that had the 'meteorite' been encountered by the Earth some five hours later, it would have blasted the city of Leningrad to oblivion, creating a devastation not unlike that later brought about by man himself in the Japanese cities of Hiroshima and Nagasaki.

The Sikhote-Alin Meteorite

On 12th February 1947, the second great Siberian meteorite of the twentieth century screamed through the atmosphere above the far eastern provinces of the USSR, and at 10.38 am

exist atomic particles with the same properties of the electron (negative electron charge) but with a positive charge called the anti-proton.

* The extraordinary events of the Tunguska object has caught the public imagination as much as did comets in ancient times. In 1969, the Russian journal *Priroda* published a catalogue of 180 scientific papers, 940 articles, 60 novels plus a whole miscellany of short stories, poems, plays, films, radio and television programmes all dealing with the Tunguska fall. Among the contributions were various explanatory hypotheses ranging from divine retribution to blast-off by spatial visitors.

local time, rained a shower of iron fragments on the western spurs of the Sikhote-Alin mountain range situated near the Chinese border (lat. 46° 9′.6N long. 134° 39′.2E).

Witnesses to the dazzling white fireball described it moving from north to south, scattering a trail of fiery sparks which blinded the eyes of the watchers and produced a series of secondary shadows in spite of a brilliant morning sun. Seconds after disappearing, a path of thick smoke extended from horizon to zenith, and loud rumblings were heard resembling artillery fire, followed a little later by roaring noises and crackling sounds like machine-gun fire.

Pilots at a nearby airfield described the fireball as "as large as the Moon" as it descended with a steep inclination to the horizon. Other witnesses spoke of it as equal in brilliance to the Sun, and a telephone technician perched on a pole said that after the flash he experienced a distinctive electric shock in spite of the wires being disconnected. The pillar of smoke remained in the sky as a reddish-coloured swirling column-cloud that persisted for hours afterwards before it was dissipated by atmospheric winds.

Three days after the fireball, some military pilots – original witnesses to the event – spotted the impact zone on the *taiga* below. From a height of 700 metres (2,300 feet) there could be no mistaking the craters which stood out like rusty yellow blobs in contrast to the dazzling white snow areas surrounding them.

A Soviet scientific expedition reached the site* towards the end of February in less than two weeks following the fall and found about 30 small craters ranging in size from 1 to 28 metres (3 to 92 feet). It was immediately apparent that no high-temperature explosive blast had occurred as with the Tunguska object. All the small craters were the result of a direct impact causing deformation of the underlying rocks and clay. Four expeditions were dispatched within 19 months of the fall; on these later ones numerous small meteoric fragments were found, and additional small impact crater-pits located. The area of craters was also mapped, showing that all

* The site lies in the wild picturesque Ussurii *taiga* where grow great cedars, walnut, ash, red birch and wild grape vines. The forest land is populated by bears, deer, boars, wapiti, racoons, badgers and tigers.

had struck the Earth in an elliptical-shaped impact zone whose major axis measured 3 kilometres.

E.L. Krinov, who visited the spot with one of the first field expeditions, returned to the area in 1967, and afterwards commented that he was amazed at the apparently unchanged appearance since 1947. Impact scars caused by large iron fragments smashing into exposed mountain rocks were excellently preserved. The only change was within the craters themselves where birch, aralia and other trees and bushes had sprung to life.

The Sikhote-Alin fall is a unique one, for it is the first purely iron meteorite shower to fall which was actually witnessed. The Russian V.G. Fesenkov, who investigated its orbit, was of the opinion that the meteoritic rain shower was represented by a single body before it broke up in the Earth's atmosphere. The orbital velocity, estimated at 14.5 kilometres per second, suggests it was once a member of the solar system in the guise of an asteroidal-type body (see also page 221) revolving in an eccentric orbit, that carried it to half way between Mars and Jupiter at aphelion and near to Venus at perihelion.

Fossil Meteorite Craters

One of the largest and oldest meteorite craters lies on the Ungava Peninsula of northern Quebec in Canada. It was first noticed as an anomalous circular lake-feature in 1950 by a prospector F.W. Chubb while scanning aerial photographs for promising rock structures. First named after Chubb, its official title has now been designated the New Quebec Crater. It measures 3,340 metres (11,020 feet) in diameter and is 360 metres (1,190 feet) deep, and the rim of the crater is raised about 100 metres (330 feet) above the surrounding terrain. Although no meteorite specimen has so far been recovered, the formation is almost certainly the result of the impact of an explosive meteorite that fell a very long time ago. The surrounding rocks consist of granite, and the local geology does not suggest that volcanic activity may have produced the crater.

With the development of intensive aerial survey as a prospecting tool, in many of the remoter parts of the world prominent circular structures have been found and critically examined to discover their origin. During the 1950s and '60s,

Canada proved to be an unusually fruitful area in the search for ancient meteorite craters, nowadays often referred to as *fossil* meteorite craters.

The term *astrobleme*, Greek for 'star-wound', is also applied to an ancient meteorite crater-structure. A criterion widely used in determining whether a suspect structure is terrestrial or nonterrestrial in origin is if a *shatter cone** is present. Many believe that shatter cones cannot be generated by the usual dynamic forces at work in the Earth's crust and can therefore only be explained in terms of a meteorite impact.

On the basis of the shatter cone criterion, over 15 fossil impact craters are now recognized in Canada ranging in size from 2.4 kilometres to 59 kilometres (1.5 to 37 miles) in diameter and in age from 100 to 500 million years (see Table page 242). Geophysical surveys reveal the presence in some of these craters of a local negative gravity anomaly. This would be expected in any structure subject to severe impact, for later the underlying rocks would expand (rebound) and become less dense. It has been suggested that an even larger ancient meteorite structure is the Nastapoka Island arc, forming the eastern shores of Hudson Bay, which is vaguely reminiscent of the Mare Crisium formation on the Moon.

Most of the suggested fossil craters lie in the belt of the ancient Canadian Shield rocks represented among the oldest Archean rocks on the Earth. Because of this it is suggested that Canada provides a reliable guide to the historical record of meteorite bombardment on the Earth's surface.

Elsewhere in the world can be found other large suspect structures. The Steinheim Basin and the Ries Kessel (Giant Kettle) feature in South Germany are two European examples. In South Africa, the Vredefort structure has been subject to a great deal of speculation. In the United States the Carolina Bays extending along the eastern seaboard have been suspected of being meteoric since 1933 after their first study by aerial photography.

* Shatter cones are rock structures of unique striated conic appearance which form under high pressure *from above*. Some geologists maintain that shatter cones may be volcanic in origin. However, in a shatter cone the mineral quartz undergoes a special tell-tale deformation which can only be explained in terms of a meteorite event.

It would seem from the recent geophysical studies of the Earth's surface that our planet during its history has been subject to intense bombardment. The vast majority of ancient craters have been obliterated through erosion and other geological evolutionary processes, but those that remain provide ample dramatic evidence to past encounters.

CHAPTER XVI

Meteoroids and Meteor Streams

The recognition of 'shooting stars' or 'falling stars' as bodies originating from cosmic space occurred about the time Chladni put forward his theory on meteorites in 1794. From the earliest times streaks or flashes of light, sometimes in the form, of spectacular displays, had been observed in the heavens, but after Aristotle had decided that they were simply a manifestation of the atmosphere* (hence the name 'meteor' from the Greek, relating to something in the atmosphere), they were ignored by astronomers until the end of the eighteenth century.

In 1798, two students, Brandes and Benzenberg at the University of Göttingen, after reading Chladni's book, decided to try the experiment of observing simultaneously the streaks of light from two locations on the ground separated by several kilometres. In spite of the crudeness of the experiment, it was an unqualified success, for their simple triangulation method showed that the light flashes occurred at heights at least 80 kilometres above the Earth's surface, and they originated from particles travelling with velocities of several kilometres per second from a source that lay beyond the region of the Moon.

The cosmic particles observed by Brandes and Benzenberg in their experiment are still known colloquially as 'shooting stars' although they are no more related to stars than are fireballs. During the nineteenth century they assumed the formal title of meteors, but by the middle of the twentieth century so much confusion resulted in discussions about particular

* As he also supposed were comets (see page 16).

cosmic particles that a precise nomenclature needed to be evolved to resolve the difficulty. Today the term *meteor* is used as a general term covering all manner of small cosmic particles, and it can be used as a noun or an adjective. The old-fashioned meteor particle is now correctly termed *meteoroid*,* although when meteoroid particles collectively form a stream, they are still ambiguously referred to as a *meteor stream*. A spectacular display of meteoroid particles is termed a *display*, *shower* or *storm* depending on the intensity of the activity. The apparent position in the sky from where a meteoroid particle enters the Earth's atmosphere is called the *meteor radiant*.

It has been estimated that several thousand million meteoroids enter the atmosphere every 24 hours. Only part of this daily influx is visible to the naked eye in the form of 'shooting' or 'falling stars'. The remainder, observed only with telescopic aids, form the considerably greater proportion. These latter particles are not to be confused with *micrometeoroids* or *micrometeorites* which are even smaller particles (see page 184).

In spite of the high number of visible meteoroids encountered by the Earth each day, a visual observer under suitable observing conditions i.e. when the sky is clear and the Moon is absent, will see only four to ten meteoroids per hour. The exact number will depend on the time of night and the season of the year. The majority of these meteoroids will be *sporadics* – or meteoroids that cannot be identified as belonging to any well-defined meteor stream.

When plotted on a star chart, the tracks of the most brilliant meteoroids appear to originate from a precise location in the heavens. This is called the meteor radiant, since owing to the effect of perspective the meteors appear to radiate outwards in all directions. Each radiant takes the name of the constellation in which it is situated† (see Table pp 244-5). The most spectacular meteor radiant of all time is positioned in the constellation of Leo (the Lion), hence these meteors are referred to as the Leonids.

* Ranging in size from a few microns up to several millimetres. A meteoroid brighter than -4^{m} is technically called a fireball even if it will not survive to be deposited on the Earth's surface as a meteorite,

† One exception is the shower known as the Quadrantids. This constellation is now obsolete and is incorporated into Boötes.

The famous naturalist Humboldt and his companion Bonpland observed the Leonids in South America in 1799 (see

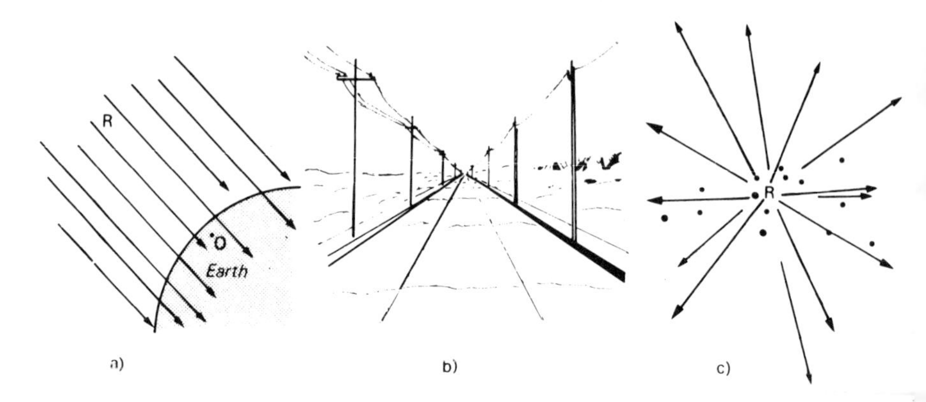

Figure 13. Meteor paths and radiants and the effects of geometrical perspective:
All meteors striking the Earth's atmosphere from a meteor stream travel in parallel paths, Figure (a), but to the observer on the Earth's surface at O, all meteors will appear to diverge from position R in the way seen in Figure (c); this is caused by the effects of geometrical perspective, and a familiar example is demonstrated in Figure (b)

Plate 48).* The pair had risen early to enjoy the morning air when purely by chance they were treated to a memorable spectacle which lasted over two hours during which time hundreds of thousands of meteors flashed across the heavens.

But Humboldt's account was not the earliest mention of the Leonids. A Turkish historian records that on 12th October AD 902, "When King Ibrahin-Ben-Ahmed died an immense number of falling stars were seen to spread themselves over the face of the sky like rain". In AD 934, the Chinese, Arabian and European chronicles mention a spectacular display about 14th October. A Japanese observation of the Leonids took place in AD 967. The account relates: "At night, on the 9th day, 9th Moon 4th year of Koh Ho, from 10 pm stars began to shoot, east to west, incessantly until 4 am appearing as glittering swords."

Another Chinese record dates to 15th October AD 1002, and

* The contemporary artists' depiction of meteors in flight is quite erroneous! (see also Plate 50).

from then on through the Middle Ages and the Renaissance the chronicles rarely failed to mention them in terms such as "Stars shot hither and thither, and flew against one another like a swarm of locusts continuing until dawn".

It was not, however, until Humboldt drew attention to it in 1799, that there was a mention that the meteors might originate from one location in the sky. This significant point was ignored until meteors became truly respectable objects a decade or so later.

The greatest *historical* display of the Leonids occurred in 1833, on the morning of 12th November, when the United States was witness to a memorable display of celestial fireworks that lived in the memories of those who saw it. Professor Thomson of Nashville afterwards wrote:

> About an hour before daylight I was called to see the falling meteors, it was the most sublime and brilliant sight I have ever witnessed. The largest [sic] of the falling bodies appeared about the size of Jupiter or Venus when brightest. The sky presented the appearance of a shower of stars and omens of dreadful events. I noticed the appearance of a *radiating point* which I conceived to be the vanishing point of straight lines as seen in perspective. *This point appeared to be stationary.* The meteors fell to the earth at an angle of about seventy-five degrees with the horizon, moving from the east towards the west!

The spectacular display was witnessed from the West Indies to Canada in the region of longitudes 60° W to 100°W. At its peak the meteors fell so thickly that it reminded one witness of a snowstorm, "and one thousand meteor flashes might be counted every minute".

A cotton planter in South Carolina, echoing the remarks of Professor Thomson, recalled his own experience of events in 1833:

> I was suddenly awakened by the most distressing cries that ever fell on my ears. Shrieks of horror and cries of mercy I could hear from most of the negroes on three plantations, amounting in all about six or eight hundred. While earnestly listening for the cause, I heard a faint voice near the door calling my name. I arose, and taking my sword, stood at the door. At this moment I heard the same voice beseeching me to rise, and saying, 'Oh, my

God! the world is on fire!' I then opened the door and it was difficult to say which excited me most – the awfulness of the scene, or the distressed cries of the negroes. Upwards of one hundred lay prostrate on the ground; some speechless, and some uttering the bitterest cries, but most with their hands raised, imploring God to save the world and them. The scene was truly awful; for never did rain fall much thicker than the meteors fell towards the earth, – east, west, north, and south, it was the same!

In 1834, two Americans, Olmsted and Twining, suggested that the annual Leonid showers were caused when the Earth passed through a cloud of meteor particles each November. Although moderate annual displays continued, none were as dramatic as the meteor storms seen in 1799 and 1833, or the previous storms noted in historical records. In 1864, H.A. Newton of Yale College, New Haven, reached the conclusion that perhaps the most dense part of the Leonid meteor cloud was only met with at intervals of approximately 33 years; this was readily apparent if one searched back into history beginning with the meteors recorded in AD 902. As a result Newton boldly predicted that the Leonids would recur as a *spectacular* display in 1866. Unknown to Newton, Dr Olbers had independently reached this conclusion some years earlier.

On the night of 13th-14th November 1866, Robert Ball,* then employed in the service of Lord Rosse, was observing with the 72-inch telescope at Birr Castle in Ireland. He afterwards related:

On that memorable evening the night was fine; the moon was absent . . . I was of course aware that a shower of meteors had been predicted, but nothing that I had heard prepared me for that splendid spectacle so soon to be unfolded . . . I shall never forget that night . . . It was about ten o'clock when an exclamation from the attendant by my side made me look up from the telescope, just in time to see a fine meteor dart across the sky. It was presently followed by another, and again by others in twos and threes, which showed that the prediction of a great shower was likely to be verified. The Earl of Rosse then joined me at the telescope and after a brief interval, we decided to cease our observation of the nebulae and ascend to the top of the wall of the great telescope, from where a clear view of the whole hemisphere of the heavens could be obtained. It was there, for the next two or three

* Afterwards Sir Robert Ball, Astronomer Royal for Ireland.

> hours, we witnessed a spectacle which can never fade from my memory.
>
> The shooting stars gradually increased in number . . . sometimes they swept over our heads, sometimes to the left, but they all diverged from the east. As the night wore on the constellation of Leo ascended above the horizon, and then the remarkable shower was disclosed. All tracks radiated from Leo. Sometimes a meteor appeared to come directly towards us, and then its path was so foreshortened that it had hardly any appreciable length, and looked like an ordinary fixed star swelling into brilliancy, and then rapidly vanishing . . . It would be impossible to say how many thousands of meteors were seen, each one of which was bright enough to have elicited a note of admiration on any ordinary night . . .

H.A. Newton's and Olbers' prediction had been amply fulfilled. During his investigations Newton had succeeded in tracing the Leonids' recurrences in a 33¼-year period beginning with the AD 902 event. This work showed that the shower had been delayed by one day in about 70 years so that it has gradually crept forward through the months of October into the middle of November. Later John Couch Adams, famous for his theoretical work in predicting the planet Neptune, succeeded in computing the orbit of the Leonid stream. Although Newton had correctly predicted the periodicity of the event as a spectacular meteor stream, his work showed that there were five possible orbits which would equally well fit the circumstances. The first of these was one of 33 years, the second a little over a year, and the third a little under a year; while there still remained a possibility of two smaller orbits. Adams' exhaustive work was a brilliant piece of theoretical research which showed that of the five orbits Newton had suggested, only the largest of 33 years would satisfy all the conditions.

During the 1860s, meteors attracted the attention of many eminent astronomers who recognized them as a new and important branch of astronomy. Apart from the Leonid meteors, it appeared there were other meteor streams that gave rise to annual showers and which likewise could be traced back in historical records. Each August occurs a display of meteors traditionally known as "the Tears of St Lawrence", so named because they coincide with the festival of St Lawrence

(10th August). These "fiery tears" have long been recorded in the church calendar of England and may be traced back to the

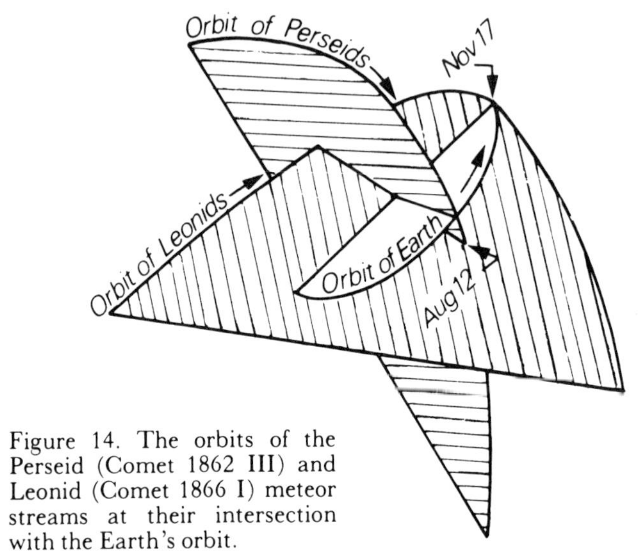

Figure 14. The orbits of the Perseid (Comet 1862 III) and Leonid (Comet 1866 I) meteor streams at their intersection with the Earth's orbit.

tenth century. Quetelet as early as 1836 showed that the August meteors were periodic. Then in 1866 the Italian Schiaparelli* unexpectedly announced that the August Perseids appeared to move in the same orbit as the periodic Comet Swift-Tuttle 1862 III. Soon after Schiaparelli's announcement the Leonid meteors were identified by Leverrier and Peters with Comet 1866 I, discovered by Tempel and Tuttle, which had a period of 33 years. Another meteor display, the Lyrids, was identified with Comet 1861 I, and the Bieliids with Biela's famous periodic comet (see page 107).

In 1861, the American Kirkwood suggested that meteors and comets may be associated in some way. He also speculated that meteors represented the fine debris thrown off from a comet which then distributes itself along the comet orbit to form a wake through which the Earth occasionally passes. This idea is still very much favoured, but there are a number of serious objections against what must be considered a rather

* Later famous for his controversial observations of Martian 'canals' in 1877.

glib and somewhat oversimplified picture of meteor streams. However, there can be no doubt that meteors and comets do have some kind of genetic association.

The meteor shower associated with Biela's Comet, known either as the Bieliids or Andromedids, is a very long standing one. Records go back to AD 524, and in recent historical times spectacular displays were witnessed in 1741, 1798 (observed by Brandes and Benzenberg), 1830 and 1838. The associated comet split in 1846 and was seen again in 1852, but never in the predicted apparitions following. The great Bieliid shower which was seen in 1872 was attributed to the final disintegration of the comet into its constituent meteors. The event was best seen in Europe, but H.A. Newton, observing from the United States in less favourable circumstances, said afterwards:

> . . . They came upon us in crowds. Over 1000 were counted in an hour, By nine o'clock the display was over . . . We saw only the last drops of a heavy shower. Before the Sun had set with us (in USA) the shooting stars were seen throughout all Europe, coming too fast to be counted. At least 50,000, perhaps 100,000 could have been seen there by a single party of observers.

Another spectacular display occurred in 1885, but since then fewer meteors have been seen. The probabilities that the 1872 and 1885 showers represented the remains of Biela's Comet is discussed elsewhere (see page 108).

In the nineteenth and early part of the twentieth centuries, observational meteor astronomy was mainly a dedicated amateur pursuit. It required little apparatus. The chief requisites were a trained eye and a great familiarity with the naked-eye stars so that meteor paths could be recorded accurately on a star chart. W.F. Denning was the greatest naked-eye meteor observer of the early period. In 1899, after more than 20 years' continuous observation, he published a catalogue of meteor radiants which itemized 4,367 separate radiants. Denning expressed the opinion that on *each* night of the year there were 50 or more meteor showers at play. Nevertheless, more than 99 per cent of the showers are minor ones and have rates of *less* than 10 meteoroids per hour. In contemporary times many of the radiants in Denning's list have been criticized, since they are considered to be based on rather slender evidence. Nowadays we know that meteor radiants are complex structures which may rapidly shift their apparent position within a few hours. Denning's list of radiants is still an important guide to the general distribution of meteor radiants, although it is certainly much less definitive than formerly thought. However, one of Denning's important minor radiants, the December Ursids, hotly disputed in the 1920s and 1930s, received independent confirmation when the Czech astronomer Bečvár rediscovered it on 22nd December 1945.

The visibility and velocity of meteors is determined by a number of factors. The time of year and the time of day are quite significant. Because the Earth moves round the Sun at a speed of 29·8 kilometres per second, some meteors are met with 'head-on' while others catch up with the Earth. Those that the Earth encounters on the 'leading side' move at high velocity (up to 72 kilometres per second), while those encountered on the sheltered side have to overtake the Earth and therefore appears to move with less velocity which may be as low as 11/12 kilometres per second. Since the apex of the Earth's way in its journey round the Sun is due south at about 6 am (page 169), more meteors will be seen about this time, ignoring the presence or absence of a major shower. It was Schiaparelli's theoretical work during the 1860s which first showed the relationship of meteor frequency as a function of the time of night, or expressed alternatively, the position of the Earth's apex in the observer's sky. It will be seen that because the apex is east of the meridian (the observer's north-south line), there will always be a preponderance of meteors from this direction (again ignoring major showers). A seasonal variation in numbers of meteors seen is brought about by the variation in the height of the apex due to the changing declination of the Sun. In the northern hemisphere more meteors will be seen in autumn (apex high in the sky) and less in spring (apex low in the sky), in the southern atmosphere the opposite will apply.

Another feature which alters the rate of meteors seen, particularly in an active shower, is the altitude of the radiant. When the radiant of a shower is just above the observer's horizon, he will see only one-tenth the number of meteors he would see if the radiant were directly overhead. This is referred to as the Zenithal Hourly Rate (or ZHR).

The appearance of a diverging meteor radiant is due to the geometrical effect of simple perspective, since all the meteoroids travel in parallel paths. The *apparent radiant* of the meteor display, as seen from the surface of the Earth is *not* the true direction in space from which the meteors are travelling in their orbit round the Sun. As a result of the combined orbital motion of the meteors and the Earth, the *true radiant* is displaced towards the Earth's apex. This effect may be illustrated by the terrestrial analogy of rain or snow falling vertically yet it appears slanted if one runs or drives through it.

Meteor streams may be perturbed by planets into new orbits in much the same way as comets. Meteor streams moving in retrograde orbits (clockwise) and/or at high inclinations to the ecliptic are least likely to be affected by short term perturbations. Most of the persistent meteor showers such as the Leonids, the Perseids (nicknamed "the Old Faithfuls") and the Lyrids either have retrograde orbits and/or high inclinations. The Lyrids have been traced back for over two and a half millennia.

Like comets, meteor orbits all show a regression of the nodes if they move in a direct orbit or an advance of the nodes if their motion is retrograde. It is this continuous shifting of the nodes which assists a computer to identify meteor particles with a particular orbit or comet, and the criterion which was used to great effect by John Couch Adams in his investigations of the Leonids.

Nevertheless, planetary perturbations may significantly influence the long-term streams. It would seem that even the long persistent streams have been captured in much the same way as comets by the dominant influence of a planet. The Leonids were captured from a very long period orbit in AD 126, after a very close approach to the planet Uranus. This had the effect of shortening the period to 33 years and also changing the orbit from a direct into a retrograde one. Between the spectacular Leonids display of 1866 and the anticipated return of the event in 1899, the main body of the stream passed close to Jupiter and Saturn. In November 1899, only a few meteors were seen, and the reason was afterwards attributed to perturbations by these major planets. The 'display' of 1933 was also disappointing, and it appeared that the spectacular Leonids had become just a part of folk-lore history. Although the annual displays continued, sometimes showing signs of intermittent activity, the Leonids had lost their former splendour. During the early 1960s, more Leonids were seen, but it was generally believed that the main body of the stream now lay well outside the Earth's orbit. Even the associated comet had become 'lost' and had not been seen since its discovery in 1866.

Nevertheless, a great surprise was in store for modern observers – leastwise in the Western United States and Eastern Siberia. On the morning of 17th November 1966, parties of optimistic observers, who had stayed up all night to catch sight of the desultory Leonid stragglers, began to notice an increase in frequency. From a rate of 20 per hour it suddenly doubled, and meteors equal to the brilliance of Jupiter and Venus began to shoot majestically across the heavens in almost uncountable numbers . . .

The sky literally began to rain shooting stars. At 4.45 am local time, one party in Arizona estimated the frequency to be several hundred per minute. A member of this party, Dennis Milton, a reporter for the popular US monthly *Sky and Telescope*, later wrote:

> . . . The view was so spectacular we just didn't know where to look! Sometimes we would spin around, taking in the whole sky. Or we alternated with looking towards the western horizon (which was very clear) and gazing right at the radiant. Different parts of the sky would light up and we would glance here and there. Everyone was yelling and laughing at the incredible dazzling sight and at our luck in seeing it!
>
> The rate was over 100 per minute for an hour, from about 4.30 to 5.30 am. It was over 1,000 per minute for 40 minutes between 4.35 to 5.15 am. The peak of perhaps 2,400 per minute was centred at 4.55 am. [Plate 51] . . . By 5.40 am the shower was back to 30 per minute. We continued to see Leonids in the brightening dawn sky until the colourful Arizona sunrise clouded out our viewing. Some said it was like a dream, an amateur astronomer's dream come true . . .

The 1966 return of the Leonids was undoubtedly the greatest meteor display ever witnessed by man since observing records began, and certainly since the Earth began to encounter the streams in historical times. It was much more spectacular than the great displays of the nineteenth century. Another surprise was the discovery of the long lost comet associated with the Leonids, P/Tempel-Tuttle, found again in 1965, albeit it was a very faint object. This comet has now been positively identified with a very bright comet seen by the Chinese in 1366. Since 1966 the annual rate has fallen off again, and each year about the middle of November only a few meteors may be seen emerging from the centre of the distinctive sickle-shape configuration of stars that forms the constellation of Leo. When the year 1999 approaches, there will be keen anticipation of another major Leonid meteor storm, but whether this will occur, is open to speculation. In spite of the precise exactness at which men may predict the apparitions of a planet, space probe or comet, this cannot be done in respect to meteor storms, for there are factors which influence their density, distribution and orbits which at the present time is not fully understood.

The Leonids are not alone in providing spectacular twentieth-century displays. In 1900, a comet (1900 III) with a period of 6½ years was discovered by the Frenchman Giacobini at Nice. When it returned in 1913, having been missed in 1906/7, it was accidentally rediscovered by the German, Zinner at Bamberg. From then on it became known as P /Giacobini-Zinner 1913 V.

At its return in October 1926, the Earth crossed the comet's orbit 70 days before the comet, and some meteors were observed which were shown to have a connection with it. These were subsequently named the October Draconids. On its return in 1933, the Earth passed the junction of the orbits 80 days later than the comet, and on 9th October a spectacular shower was seen which reached a peak rate of 300 to 1,000 meteors per minute.

In 1939, the Earth was 136 days ahead of the comet and no meteors were recorded, but in the 1946 return the Earth crossed the comet's orbit only 15 days behind, and a magnificent shower was witnessed. This was the first great meteor shower to be observed using a new radio technique (see below).

Since 1946, no meteor shower connected with P/Giacobini-Zinner has been seen.* For some years the ever-present perturbing influence of Jupiter lessened the perihelion distance so that for a number of apparitions the streams moved *inside* the radius of the Earth's orbit as it crossed its path. However, in 1969, the comet's orbit made a close approach to within 0·577 AU of Jupiter which had the effect of switching it back towards the Earth. The comet returned in 1972, when the Earth lay within 0·00016 AU distance of the orbit 58 days after the comet passed by, but although optimistic predictions were made for a spectacular meteor shower to recur on 8th October 1972 at 15.45 Universal Time, very few Draconid meteoroids were seen. However, about this time radar observations in Japan showed a peak of 84 return echoes during one 10-minute period, and the average rate over the whole day was 20–30 echo returns per 10-minute interval.

The Draconid meteor stream is a very narrow one in comparison to some of the older, well-established streams. In 1933 and 1946 the Earth ploughed through it in about 6 hours, indicating a cross-sectional width of 640,000 kilometres (400,000 miles). In contrast, the Perseid stream is much wider, and it takes the Earth nearly two weeks to pass through it, so that its real width cannot be less than 80 million kilometres (50 million miles).

Among the lesser annual showers the Geminids, 11th–13th December, are of great interest. The orbits of the meteoroid particles revolve in a period of 1·65 years and approach the Sun to within 0·14 AU. It is the shortest period of any known meteor stream, and if it is associated with an extinct comet, it is difficult to conceive how such an object was perturbed into an orbit of this kind. The Geminids are distinctive in that the meteoroids which form the annual display appear to be extraordinary dense particles of matter quite different from the fragile wisps of cosmic material that seems to comprise most meteor streams.

* As of 1972.

44. An australite; the distinctive button shape is due to aero-dynamic modelling during flight.

45. Tektites take many forms; they can be walnut-sized (or smaller) or as big as an apple, and they range in colour from black through brown to a kind of bottle green.

46. The Tunguska taiga (USSR), scene of the fall of an unusual meteorite on 30th June 1908.

47. Sections through stony meteorites (chondrites) showing the characteristic circular chondrules (all scale bars = 0.2 mm).

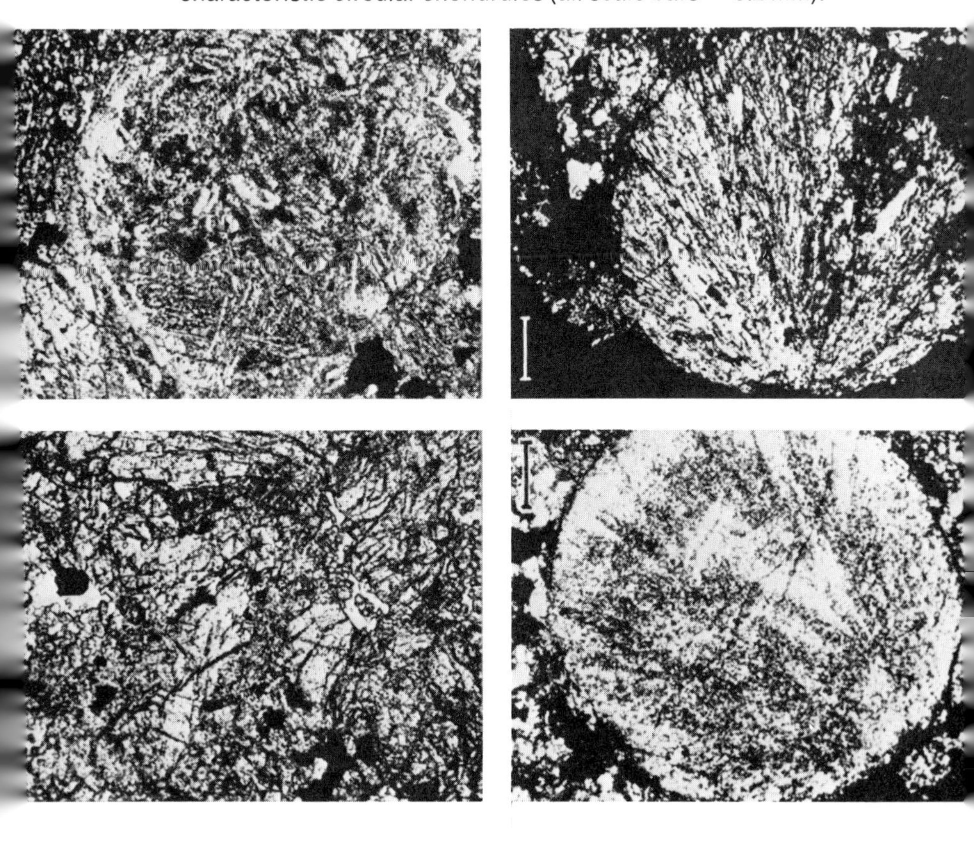

48. Bonpland and Humboldt observing the Leonids in South America in 1799.

49. The Hoba West meteorite. This 60-ton mass of iron lies where it was found in South-West Africa.

50. An old print depicting the Leonid meteor shower in 1833.

51. The Leonids in 1966.

From the vivid and dramatic descriptions recorded by eyewitnesses, the impression is conveyed that meteor streams, giving rise to meteor storms, consist of dense aggregations of cosmic material. By *average* cosmic space densities this is indeed the case. However, in comparison with terrestrial standards the mean density of a meteor stream is very low. For example, the spatial distance between *each separate* meteoroid particle* in the great Leonid display of 1966 was about 15 kilometres. The Perseid particles are separated by 300 kilometres while the sporadic non-shower meteors by 650 kilometres. These rather surprising figures help exemplify the very emptiness of cosmic space immediately outside the terrestrial atmosphere.

Modern professional meteor observation dates from about 1936, when the Harvard College Observatory began a programme of meteor photography to determine meteor speeds using short-focal-length, wide-angle cameras. In 1952, these cameras were replaced by ultra-fast Baker Nunn Super-Schmidts which operate at *f*/0·67. With these cameras it is possible to record all meteors which would be visible as naked-eye objects, and they have assisted in accurate determinations of meteoroid particle orbits prior to atmospheric burn-up.

During World War II, it was accidentally discovered that meteor trails could be detected in the atmosphere with radar. The meteor trail produces a temporary wake of ionization from which the radar beam is reflected back to the receiver, revealing its presence in the atmosphere. When hostilities were over, war surplus apparatus was modified for observing meteors, and a whole new technique of observing meteors was evolved, independent of weather and time of day. New important annual meteor showers were detected which only occur in daylight hours (see Appendix VII), and one of them, the Beta Taurids, active between 24th June–5th July, is probably associated with Encke's Comet.

The introduction of radar observing techniques has shown that meteor radiants are sometimes very complex structures. Some individual radiants appear to overlap and may shift position within a few hours. Since continuous and systematic

* Each probably smaller than a grain of sand.

observations became possible throughout the 24 hours, the number of known major showers has almost doubled.

CHAPTER XVII

Origin of Meteoroids and Meteorites

The origin and nature of meteoroid particles and the larger meteorites is one of the most interesting and perplexing problems in planetary science. There are fundamental questions that have not yet been satisfactorily answered such as: Do meteoroids originate from interstellar space or are they original constituent members of the solar system? Do meteoroids owe their origin to cometary or asteroidal decay, or both? And following that question: Are meteoroids, micrometeoroids, micrometeorites, meteorites, comets, asteroids, and the interplanetary 'solid' medium which forms the zodiacal light, interrelated genetically?

From the orbital/dynamical point of view there is convincing evidence to show that at least comets and meteor streams (showers) are closely related in some way. Although the work carried out by the wide-angle Super-Schmidt sky cameras shows that only 30 per cent of all meteoroids can be classified as belonging to one of the established meteor showers, it is likely that the remaining 70 per cent forming the sporadics belong to minor showers that have become highly dispersed owing to long-term planetary perturbations.

In the early days of meteor observing it was thought that 'shooting stars' possessed hyperbolic velocities indicating they originated from outside the solar system and were merely passing the Sun when encountered by the Earth. Modern velocity measurements have shown that this earlier assumption was wrong. Although there are still a few authenticated instances of hyperbolic meteor orbits, they represent very borderline cases and are the result of local planetary perturbations in much the same way as the hyperbolic perturbations induced in comets in near encounters (see page 52). The evidence of the origin of most present-day meteoroids from within the solar system is overwhelming, yet in the past the situation may have been different. The presence of meteoroids can be satisfactorily (if somewhat glibly) accounted for either by assuming they represent cometary debris left in the comet's wake from a 'dirty snowball' comet model object (see page 59), or that they are associated with comets in similar orbits yet remain independent objects. At the present time it is not possible to say which idea is correct, but most astronomers would be inclined to accept the first one. However, it might well be that *both* ideas are correct. If comets form via the Lyttleton accretion axes hypothesis (the 'sand bank' model), in which the Sun's gravitational lens focuses captured interstellar particles which then collect or coalesce to form the head of a comet (see page 61), some may remain unattached yet continue to follow the elliptical orbits impressed on them by the Sun. Certainly not all meteoroid particles which the Earth encounters, or gravitationally attracts, follow *behind* in the orbital wake of the comet. The Bieliid meteors in particular appear randomly bunched in various parts of their orbit, and the October Draconid meteors have provided magnificent displays *ahead* of the comet. If meteors are simply the result of cometary decay, they would eventually surround the entire comet orbit in an even distribution, analogous to a kind of continuous bicycle tube, a situation which is not evidenced by present-day observations.

The position is complicated by the fact that concentrations of meteoroids located in various parts of an orbit will be perturbed into different orbits if they encounter the near approach of a planet; in this way an original meteor stream will be divided into several orbits of different, yet related, orbital elements. This is most certainly the case with the Bieliid meteors and also with the daytime Taurid meteors probably associated with Encke's Comet.

The ultimate fate of a small meteoroid particle in space has a number of alternatives. There are several different mechanisms by which a meteoroid particle can be destroyed or removed from its orbit. They include planetary collision, physical erosion in space due to solar corpuscular sputtering, and the Poynting-Robertson (P-R) effect which many consider to be the most important mechanism of all. This effect was first suggested by Poynting in 1903, and then given a firmer mathematical basis by Robertson in 1937, who proved it as a consequence of Einstein's theory of relativity. It was shown that when a small particle in space absorbs heat via solar radiation, the re-emission of some of the energy creates a tangential drag, or retarding force, which is proportional to the velocity of the particle. As a result of this continuous action a small particle such as a meteoroid will eventually spiral in towards the Sun and be absorbed. If indeed the P-R effect is a real one, it means that *all* the small particles moving in orbits about the Sun will eventually be swept up in a relatively short time of 100,000 years – a very short time indeed on the cosmic time-scale. However, there are two factors which may alter this. The P-R effect can only be significant on truly spherical particles of matter, and there is good reason to suppose that much micro-cosmic matter is highly irregular in structure. The variable nature of the solar wind may also act contrary to the P-R effect. At sunspot maximum there may be a 'pushing away' or 'spiralling out' of material further into space owing to solar radiation pressure, while at sunspot minimum, when the solar wind drops down to the strength of a 'breeze', the material would change direction and spiral back inwards. Thus small meteoroid particles could be shuttled back and forth over the 11-year duration of the sunspot cycle and so remain in space much longer than the P-R effect predicts.

On the evidence provided by their orbits and on other dynamical considerations, there appears to be no close relationship between *meteoroids* and *meteorites*. Until a meteorite fell in the town of Přibram, near Prague in 1959, there was a little definitive information to show the nature of meteorite orbits prior to their entry into the Earth's atmosphere. The spectacular Přibram, fireball, burning brilliantly at -19^{m}, was photographed by network cameras engaged on a systematic survey of faint meteors in the upper atmosphere. By a lucky chance it was recorded at two stations 40 kilometres apart so that from this baseline it was possible to compute with great accuracy its pre-entry orbit. After descent, 17 fragments were found on the ground, and for the first time it was possible to associate a recovered meteorite with an accurate orbit determination. This orbit showed that prior to time of collision with the Earth, the meteorite revolved round the Sun with a perihelion distance near to the orbit of Venus and an aphelion distance near to Jupiter. In other words it had all the characteristics of a typical small Earth orbit-crossing asteroid such as Icarus which passed the Earth within 6.8 million kilometres (4.25 million miles) in 1968. The age of the Příbram meteorite was determined using various dating methods (see page 185) which showed it to have formed 4 aeons ago – approximately the same age as the Earth (4·6 aeons).

Since the Příbram stony meteorite fell in 1959, a similar event occurred in the United States where the Prairie Network cameras, operated by the Smithsonian Institution, succeeded in trapping a brilliant fireball which afterwards dropped a meteorite near Lost City, Oklahoma. The finding of the first fragment was a remarkable story in itself. After the fireball pictures had been processed, a prediction was made where the resulting meteorite might be found on the ground, and it was actually located 600 metres from the very spot! The orbit showed that prior to its atmospheric entry it had the elements of an Earth orbit-crossing asteroid, but the shape was less eccentric than that of the Příbram object (see Figure 15), while its estimated age was about the same.

The study of the origin and nature of meteorites provides important information regarding the origin of the solar system. At one time it was believed that meteorites represented primary matter which had remained unaltered from the time matter had formed in the solar system. However, the study of the chemistry and general morphology of meteorite specimens show them to be highly complex bodies which have evolved into a great diversity of types and subtypes during their long lifetimes in space. Apart from the more definitive evidence provided by the Příbram and Lost City meteorites, there is additional evidence to indicate they all have orbital characteristics in common with asteroids. The classical theory that accounts for the asteroids is Olbers' (hypothetical) planet or planets (see page 77) which supposedly disintegrated or collided in the dim past and provided the necessary fragments;

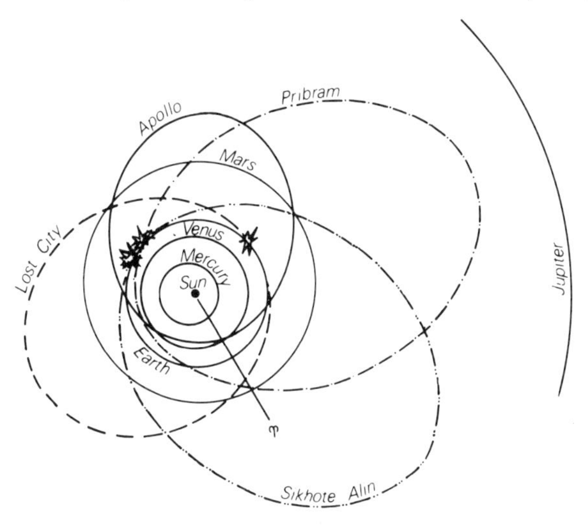

Figure 15. The orbits of the Sikhote Alin, Přibram and Lost City meteorites before they collided with the Earth. Note the elliptical orbit of the Earth-orbit crossing the asteroid (minor planet) Apollo, which resembles those of the meteorites.

and this same theory will also readily account for the smaller meteorite fragments. An alternative idea suggests that there had never been a large planet between Mars and Jupiter, but there was instead a numerous population of proto-planets, sometimes referred to as planetesimals, which owing to their mutual and constant collisions have given rise to all the necessary interplanetary debris – in a diminishing size order range – that could well account for all the diversity of coarse and fine material (asteroids, micrometeorites, meteorites, and the solid interplanetary material which gives rise to the zodiacal light) except for 'cometary' meteors.

Another idea first put forward in the middle of the nineteenth century connects part of the asteroid population with comets. This idea has received further support in more contemporary times. Two comets, P/Arend-Rigaux and P

/Neujmin 1 were both nebulous-looking objects when first discovered; in recent apparitions, however, both have shown sharp stellar-like images which had not been noted previously. Had such comets been so observed when first discovered, they would automatically have been included among the asteroids. Hind and John Herschel in the 1850s reported on the variable and often fuzzy appearance of the asteroid Irene, discovered by Hind in 1851. Since that time it has always appeared quite stellar.

Among the various classes of meteorites that have been subject to intensive laboratory examination are the puzzling and controversial carbonaceous chondrites – a rare variety of stony meteorite. Only about thirty examples are known, all of which were collected within a short time of their falling to Earth. It is generally assumed that many more specimens have fallen, but they go unrecognized owing to their inherent very friable nature and the presence of water soluble compounds that cause them to rapidly disintegrate within a few days of landing. Part of one such example recovered after falling was placed in a glass of water where it rapidly dissolved to give off a most unpleasant stink!

Carbonaceous chondrites differ from ordinary stony chondrites – to which they belong as three subtypes – for they contain hydrocarbons, water soluble salts, and most significantly a multitude of organic-like compounds. Type I contains the greatest amount of water and 'organic' matter, with only traces of high temperature minerals. Type III contain the least amount of water and 'organic' matter and have an abundance of high temperature minerals plus metallic constituents. Type II are intermediate specimens between I and III. Usually they are only recovered in small pieces weighing a few kilogrammes owing to their fragile structure. During their turbulent descent through the atmosphere, repeated thermal shocks quickly shatter any large cosmic lumps. Some specimens contain veins of magnesium sulphate indicating that they might possibly have been deposited from water solutions – evidence that liquid water may once have existed in the parent body. The micro-structures reveal a large variety of complex 'organized' molecules. Modern analyses suggest that they could have originated in any of three ways: abiological, extraterrestrial or by terrestrial biological (contamination) processes. Arguments put forward by different research workers to champion the various radical ideas have involved the whole subject in controversy for the past 150 years.

One of the most famous examples of carbonaceous chondrites (Type I) fell at Orgueil, France, on 14th May 1864. Probably no other fragment of any meteorite has been subject to such intensive and repeated study as the Orgueil specimen; and no other meteorite has been subject to such heated debate. It is agreed that as the parent body evolved, it had three main stages of mineral formation. First it was subject to temperatures of several hundred degrees centigrade which was followed by a middle stage below 170°C and then by a final stage of below 50°C. During its complicated history, somewhere in cosmic space, it was repeatedly fragmented, recemented and leached by aqueous solutions. The chemical history of the Orgueil fragments is one of the most puzzling features about the whole body. It is simply not possible to reconstruct a non-contradictory step-by-step development – as one might do with terrestrial rocks – and suggest a process by which its petrological nature came about.

In 1868, four years after the Orgueil meteorite fell, the French investigator M. Berthelot announced he had isolated saturated hydrocarbons from the fragments and thought them to be comparable with those in petroleum. This announcement was given wide publicity, for here was a possible link between life on Earth and elsewhere in the Universe. But Bethelot's findings were not the first to suggest extraterrestrial life-forms. In 1834, J. J. Berzelius described his extraction (with water and by distillation) of a whole complex of substances from the Alais meteorite, another Type I carbonaceous chondrite which fell at Alais in France on 15th March 1806. Berzelius commented in his report:

> ... There can thus be no doubt that the stone under investigation despite all its external differences, is a meteorite, which in all probability hailed from the usual home of meteorites ... Does this carbonaceous earth indeed contain humus or a trace of other organic compounds? Could this give a hint as to the presence of organic formations on other planets?

In 1859, two well-known chemists, M. F. Wöhler and M. Hornes, following Berzelius' lead, extracted organic substances (using alcohol) from the Kaba carbonaceous meteorite which fell at Kaba in Hungary on 15th April 1857. They announced that these were of humic of bituminous matter.

Many modern investigators have for the most part rejected all these early identifications of organic material. They rightly pointed out that analytical techniques were then still in their infancy, and what methods were used were crude and inconclusive. Nevertheless, the idea of genuine organic material as a constituent element of carbonaceous chondrites has persisted into the modern era.

In 1961, B. Nagy, a Hungarian working at Columbia University, and Claus, his associate, caused great excitement with their identity of *organized elements* in the Orgueil meteorite which raised such a controversial dust that it divided expert laboratory meteoriticists into two camps.

The problem of proving the existence of organized elements is one involved with terrestrial contamination. There is no doubt that hydrocarbon substances extracted from carbonaceous fragments are real enough, but it is the suggestion that they contain organized material akin to life-forms which some geochemists are reluctant to accept. It is impossible to prevent terrestrial contamination in any meteorite which falls to the surface of the Earth. The very act of passing through the terrestrial atmosphere is sufficient to expose it to contamination. Many of the lunar rock samples recovered during the first Apollo missions were subject to terrestrial contamination in spite of the ultra-clinical precautions taken beforehand. Most of the nineteenth-century carbonaceous chondrites have been stored on museum shelves for over a century with little thought to cleanliness. It is not surprising that some geochemists are sceptical about the finding of life-forms in them in more modern times!

However, the introduction of scanning microscopy began a new era in the examination of carbonaceous chondrite specimens. It allows for a critical examination of the microstructures, because it permits the inspection of surfaces at high magnification and at higher than usual focal depths. Scans of freshly broken surface of the Orgueil specimens have revealed that organized elements are partially embedded in them. In view of this it would appear that these particular organized elements are not terrestrially contaminated. But the doubting Thomases of meteoritics are still highly sceptical – especially about an idea that some so-called organized material represents extraterrestrial microfossils.

In 1969, the British chemists J. Brooks and G. Shaw announced that they had detected an unusual chemical in a fragment of the Orgueil meteorite* which is almost certainly not a contaminant, since it forms some 4 per cent of the meteorite's weight. They identified it as sporopollenin, a complex biological material that forms the outer coat of pollen grains. Using various tests they showed that this sporopollenin material resembles sporopollenin in fossil and present-day plants. The difficulty is that the exact chemical structure of sporopollenin is not known, and this is a stumbling block to the general acceptance of the findings that this organized material is extraterrestrial. Shaw has elaborated further on his ideas about extraterrestrial life. He believes that among the ancient rocks of the Earth we should already have found traces of the prebiotic soup, if life evolved on Earth through natural processes. He also believes that the evidence so far presented should be sufficient to convince the sceptics that life was seeded on the primitive Earth from outer space.

The problems surrounding the origin of life on Earth, and the possible role of carbonaceous chondrites in the propagation of life, need to take into account the remarkable experiments of S.L. Miller in the USA under the direction of H.C. Urey in the early 1950s. They conceived the idea of mixing together the gases ammonia, methane, water vapour and hydrogen which were assumed to be abundant in the primitive atmosphere of the Earth. The mixture was circulated through an electric discharge, and at the end of a week it was discovered that the water contained several types of amino acids. The experiment indicated that life on Earth might have occurred by the interaction of electrical energy with elementary chemo-biological material. As a further consequence of this, the idea has been put forward that similar primitive chemo-biological material could equally well be simulated in

* And also in one found at Murray, Kentucky in 1950.

carbonaceous material in orbit if exposed to long term intense cosmic radiation without the necessity of invoking the idea of 'life'.

A more recent example of a Type II carbonaceous chondrite fell on 28th September 1969, in Victoria, Australia. At 11 am a local dairy farmer emerged from his house to see an enormous ball of fire sweeping across the sky, which he described as producing "a crackling noise like a big heap of twigs". Hundreds of others in the surrounding countryside were witness to the fireball which finally exploded and broke up near the town of Murchison, scattering its stony debris over an elliptical-shaped area 8 kilometres (5 miles) long by 1.5 kilometres (0.9 miles) wide. In the two weeks following, 5 kilogrammes of fragments were recovered and identified as an intermediate Type II carbonaceous chondrite.

Specimens sent to the NASA Ames Research centre in California were quickly analysed, using new methods. The results revealed an abundance of 5 of the 20 amino acids normally found in living cells. Along with these were found lesser amounts of 11 other amino acids which are compounds that are structurally almost identical to protein-forming amino acids, but which have no functional role in living organisms. These amino acids have right hand and left hand molecular structures which can be easily recognized by their optical properties. Similar amino acids on Earth of biological origin are all left-handed which if interpreted correctly, implies that the mixed amino acids found in meteorites are *not* of biological origin. Nevertheless, in addition to the amino acids, hydrocarbon compounds were isolated, these being a mixture of *both* biological and non-biological types. Although not all laboratory research workers will accept these findings uncritically, it now at least appears that there is very strong evidence for the idea that meteorites show indications of extraterrestrial chemical evolution of a kind that precedes the origin of life.

Meanwhile, the controversy continues, but the Nobel Laureate, H.C. Urey, after a long critical evaluation of the whole subject succinctly summed it up " . . . if found in terrestrial objects some substances in meteorites would be regarded as undisputedly biological".

Putting aside the arguments for or against organized elements, carbonaceous chondrites, as stony cosmic bodies, pose other problems about their origin and distribution. If they originate from asteroids or planetesimals, as their complex petro-history suggests, such material might originate from a surface layer of a parent body before it suffered fragmentation into meteoric lumps. A lunar origin was once suggested, but this idea now seems very unlikely, and no rock so far recovered from the lunar surface has shown any resemblance to carbonaceous material.

The significant fact that carbonaceous chondrites are very friable structures has not gone unnoticed by those interested in the origin of comets and meteoroids. It has been speculated that the nature of such material fits the now classical ideas of fragile, low-density meteoroid 'dust balls' which are suggested by the physical behaviour of 'shooting stars' when they hit the atmosphere of the Earth. At one time it was believed that most super-brilliant fireballs dropped a meteorite of some kind. Nowadays this idea has been challenged, for the number of meteorites found in relation to the number of super-brilliant fireballs observed is very low. It would seem that many fireballs are completely dissipated in the atmosphere, and material which is able to penetrate to the surface of the Earth must possess high structural strength such as the tougher chondrites and irons. Carbonaceous chondrites may thus form a substantial part of the brilliant fireball phenomena, but owing to their inherent structural weakness, few specimens are recovered on the surface. After the Revelstoke carbonaceous meteorite (Type I) fell in Canada on 31st March 1965, only *one gramme* of meteorite was found on the surface of the Earth. However, high-flying jet aircraft of the US Air Force 9th Weather Reconnaissance Wing recovered large quantities in the atmosphere through which the fireball had ploughed, representing many tons of meteoric debris. This and other similar corroborating evidence indicates that the terrestrial atmosphere had a marked selection effect, and that museum collections of recovered meteorites are *not* truly representative of the meteorite population at large in cosmic space.

The question relating to the possible association of carbonaceous meteorites with cometary heads is of great interest. If comet nuclei exist as discrete bodies, then bodies like carbonaceous chondrites rich in carbon and associated with ices would be in keeping with the spectroscopic evidence provided by comets; and the same results would be observed if the carbonaceous chondrites were aggregations – in keeping with the 'sandbank' formation idea. Equally significant is the fact that carbonaceous chondrites yield up large quantities of gas. The Cold Bokkeveld carbonaceous chondrite specimen which fell in South Africa on 13th October 1838, has yielded very large quantities of CO_2 in relation to its volume.

Nevertheless, such speculations fail to take note that carbonaceous meteorites are far from being simple agglomerations of elementary matter. If one attempts to explain the highly complex rock structures, including the secondary mineralization plus a variety of hydrocarbons, fatty and aromatic acids, and the polymer-like substances in terms of cometary particles or a discrete nucleus, the ideas appear irreconcilable. But the idea that the Apollo type Earth-orbit-crossing asteroids represent the cores of inactive comets is now gaining a few adherents. However, on the present evidence it seems that carbonaceous chondrites originated in large monolithic bodies such as planets, or asteroids and were formed in an environment not too dissimilar to the Earth's.

One of the most puzzling features of the ordinary stone chondrites is how their chondrules formed, and how they became located in position, for they are quite unlike terrestrial rocks. Most speculative ideas consider that the chondrites are solidified droplets of once molten matter. Their crystal structures are enigmatic, for instead of radiating from a chondrule centre as one might expect, they radiate eccentrically from a point outside the chondrule (Plate 47). It seems that only a highly complicated chain of events can satisfactorily explain their evolution. They certainly show signs of a repeated cycle of heating and cooling, and in some their brecciated nature, with its consequent hotch-potch of mineralization, defies rational explanation without knowing the nature of the parent body from which it evolved.

Meanwhile, until the space probes of the future bring back a drill core of asteroidal material and reveal the enigmatic secrets of the comet nucleus, we must wait more definitive answers. Yet, theories based on slender fact, colourful conjecture and idle speculation spawn unabated. Perhaps from time to time we need reminding of that piece of cautionary scientific advice known as Occam's Razor – first set down by William of Occam (1300–1349) – which states: *Neither more, nor more onerous, causes are to be assumed than are necessary to account for the phenomena.*

APPENDIX I

Kepler's Laws

(1) The orbit of each planet is an ellipse having the Sun in one of its foci.

(2) The motion of each planet in its orbit is such that the radius vector from the Sun to the planet describes equal areas in equal times.

(3) The squares of the periods in which the planets describe their orbits are proportional to the cubes of their mean distance from the Sun.

Newton's Laws

(1) A body remains at rest or continues to move in the same straight line with constant speed unless it is acted upon by another force.

(2) The force applied to a body is in the direction of the acceleration imparted to the body, and is equal to the mass of the body times its acceleration.

(3) Every action has an equal and opposite reaction.

APPENDIX II

The Orbit of a Comet/Meteorite/Meteoroid

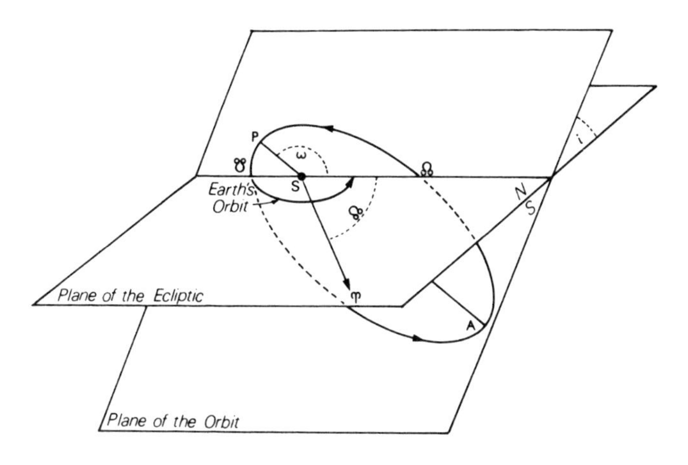

Figure 16. The orbit of a comet/meteorite/meteoroid

S = Sun
♈ = First point of Aries.
P = Perihelion (nearest point of the orbit to the Sun).
A = Aphelion (furthest point of the orbit from the Sun).
ω = Argument of perihelion (the angle from the ascending node to the perihelion measured in the plane of the ecliptic).
☊ = Ascending node (point where the orbit intersects the plane of the ecliptic, the comet/meteorite/meteoroid moving northwards (N)).
☋ = Descending node (point where the orbit intersects the plane of the ecliptic, the comet/meteorite/meteoroid moving southwards (S)).

i = Inclination of the orbit to the ecliptic plane.
Orbital elements required to determine the orbit of a comet/meteorite /meteoroid.
Time of perihelion passage (T).
Period (P) in Gaussian years = 365·25 days.
Eccentricity (e) of the orbit.
Perihelion distance (q) in astronomical units (AU).*
Argument of perihelion (ω).
Longitude of ascending node (Ω).
Inclination (i). An angle between 0° and 180°; the motion is direct (D) if the inclination is *less* than 90° and retrograde (R) if the inclination is *greater* than 90°.
Longitude of perihelion ($\tilde{\omega}$ or π) equals $\Omega + \omega$. This element is now obsolete and is only retained for statistical comparisons.

BIBLIOGRAPHICAL NOTE:

For the general reader seeking a comprehensive introductory guide to historical and contemporary astronomy, see the author's *Astronomy in Colour* (Blandford Press, London; Macmillan, New York, 1972). See also the author's *What Star is That?* (Thames & Hudson, London; Viking Press, New York, 1971). Both these books contain star maps which depict the principal meteor stream radiants.

* (1 AU = Earth-Sun unit distance = 149,500,000km).

APPENDIX III

Short Periodic Comets Observed At More Than One Apparition

Approximate elements are given for the most recent perihelion passages.

Name	P(yrs)	e	q	ω °	Ω °	i °
Encke	3·30	0·847	0·339	185·9	334·2	12·0
Grigg-Skjellerup	4·91	0·703	0·858	356·4	215·4	17·6
Honda-Mrkos-Pajdusakova	5·21	0·815	0·556	184·2	233·1	13·2
Tempel (2)	5·26	0·549	1·364	191·0	119·3	12·5
Neujmin (2)	5·43	0·567	1·338	193·7	328·0	10·6
Brorsen	5·46	0·810	0·590	14·9	102·3	29·4
Tuttle-Giacobini-Kresak	5·48	0·639	1·123	38·0	165·6	13·8
Tempel-Swift	5·68	0·638	1·153	113·6	290·9	5·4
Tempel (1)	5·98	0·463	1·771	159·5	79·7	9·8
Pons-Winnecke	6·30	0·639	1·230	172·0	92·9	22·3
de Vico-Swift	6·31	0·524	1·624	325·4	24·4	3·6
Kopff	6·31	0·555	1·520	161·9	120·9	4·7
Giacobini-Zinner	6·41	0·729	0·934	172·9	196·0	30·9

Name	P(yrs)	e	q	ω °	☊ °	i °
Forbes	6·42	0·553	1·545	259·7	25·4	4·6
Schwassmann-Wachmann (2)	6·53	0·383	2·157	357·7	126·0	3·7
Wolf-Harrington	6·54	0·538	1·614	187·0	254·2	18·5
Biela	6·62	0·756	0·861	223·2	247·3	12·6
Wirtanen	6·67	0·543	1·618	343·5	86·5	13·4
d'Arrest	6·67	0·614	1·369	174·5	143·6	18·1
Perrine-Mrkos	6·71	0·643	1·271	166·0	240·2	17·8
Reinmuth (2)	6·71	0·457	1·932	45·5	296·2	7·0
Brooks (2)	6·72	0·505	1·763	197·1	176·9	5·6
Harrington	6·80	0·559	1·582	232·8	119·2	8·7
Arend-Rigaux	6·82	0·600	1·437	328·9	121·6	17·8
Johnson	6·86	0·377	2·247	206·0	118·2	13·9
Finlay	6·90	0·703	1·077	321·6	42·1	3·6
Borrelly	7·02	0·604	1·452	350·8	76·2	31·1
Daniel	7·09	0·550	1·661	10·8	68·5	20·1
Harrington-Abell	7·22	0·522	1·785	338·2	146·0	16·8
Holmes	7·35	0·379	2·347	21·8	329·6	19·5
Faye	7·38	0·576	1·608	203·6	199·1	9·1
Whipple	7·46	0·353	2·471	190·0	188·4	10·2
Ashbrook-Jackson	7·49	0·396	2·314	349·0	2·3	12·5
Reinmuth (1)	7·60	0·487	1·983	9·4	121·2	8·3

Name	P(yrs)	e	q	ω °	☊ °	i °
Arend	7·79	0·534	1·832	44·5	357·6	21·7
Oterma	7·88	0·144	3·388	354·9	155·1	4·0
Schaumasse	8·18	0·705	1·196	52·0	86·2	12·0
Wolf	8·43	0·395	2·507	161·1	203·9	27·3
Comas Sola	8·59	0·576	1·777	40·0	62·8	13·4
Vaisala (1)	10·46	0·636	1·741	44·2	135·4	11·3
Neujmin (3)	10·95	0·588	2·032	144·8	156·2	3·8
Gale	10·99	0·761	1·183	209·1	67·3	11·7
van Biesbroeck	12·41	0·550	2·409	134·2	148·8	6·6
Tuttle	13·61	0·821	1·023	207·0	269·8	54·7
Schwassmann-Wachmann (1)	16·10	0·132	5·538	355·8	321·6	9·5
Neujmin (1)	17·97	0·774	1·547	346·7	347·2	15·0
Crommelin	27·87	0·919	0·743	196·1	250·4	28·9
Tempel-Tuttle	32·91	0·904	0·981	172·6	234·4	162·7
Stephan-Oterma	38·96	0·861	1·596	358·4	78·6	17·9
Westphal	61·73	0·920	1·254	57·1	347·3	40·9
Brorsen-Metcalf	69·06	0·971	0·485	129·5	311·2	19·2
Olbers	69·57	0·930	1·179	64·6	85·4	44·6
Pons-Brooks	70·86	0·955	0·774	199·0	255·2	74·2
Halley	76·04	0·967	0·587	111·7	57·8	162·2
Herschel-Rigollet	156·0	0·974	0·748	29·3	355·3	64·2
Grigg-Mellish	164·3	0·969	0·923	328·4	189·8	109·8

APPENDIX IV

Comets of Long or Indeterminate Period

(Where the period is given this can only be considered a tentative figure).

Name & Year	P(yrs)	e	q	ω °	☊ °	i °
'Brilliant Comet' 1264	–	1·0	0·824	159·7	150·4	16·4
Brahe 1577	–	1·0	0·178	255·7	30·5	104·9
Hevelius 1664	–	1·0	1·026	310·7	85·3	158·7
Hevelius 1665	–	1·0	0·106	156·1	232·0	103·9
Kirch 1680	–	0·9	0·006	350·6	275·9	60·7
Sarabat 1729	–	1·0	4·051	10·4	313·6	77·1
Klinkenberg	–	1·0	0·222	151·5	48·6	47·1
Flaugergues 1811 I	–	0·9	1·035	65·4	142·3	106·0
Donati 1858 VI	–	0·9	0·578	129·1	166·6	117·0
Tebbutt 1861 II	–	0·9	0·822	330·1	280·2	85·4
Coggia 1874 III	–	0·9	0·676	152·4	119·8	66·3
Wells 1882 I	–	0·9	0·060	208·9	205·9	73·8
Daniel 1907 IV	–	0·9	0·512	294·5	143·6	8·9

Name & Year	P(yrs)	e	q	ω °	Ω °	i °
Morehouse 1908 III	–	1·0	0·945	171·6	103·8	140·2
'Daylight Comet' 1901 I	–	0·9	0·129	320·9	89·3	138·8
Delavan 1914 V	–	1·0	1·104	97·5	59·7	68·0
Stearns 1827 IV	–	0·9	3·683	11·1	214·9	87·7
Skjellerup 1927 IX	–	1·0	0·176	47·2	77·5	85·1
Whipple-Fedtke-Tevzadze 1943 I	–	0·9	0·992	39·8	100·1	19·7
'Southern Comet' 1947 XII	–	1·0	0·110	196·2	336·6	138·5
'Eclipse Comet' 1948 XI	–	0·9	0·135	107·3	210·3	23·1
Pajdusakova 1954 II	–	1·0	0·072	94·1	114·6	13·6
Arend-Roland 1957 III	–	1·0	0·316	308·8	215·2	120·
Mrkos 1957 V	12800	0·9	0·355	40·3	67·6	93·9
Alcock 1959f	–	1·0	0·165	300·6	225·1	108·0
Alcock 1959 IV	–	1·0	1·150	124·6	159·2	48·3
Wilson-Hubbard 1961 V	–	1·0	0·040	270·6	298·3	24·2
Humason 1962 VIII	2900	0·9	2·133	233·5	154·7	153·3
Alcock 1963 III	15400	0·9	1·537	146·6	42·7	86·2
Ikeya 1964 VIII	363	0·9	0·822	290·8	269·3	171·9
Everhart 1964 IX	6690	0·9	1·259	20·7	279·7	68·0
Alcock 1965 IX	–	1·0	1·293	150·5	174·9	65·0
Barbon 1966 II	34000	0·9	2·019	136·2	167·0	28·7
Ikeya-Everhart 1966 IV	1830	0·9	0·879	49·8	106·8	48·3

Name & Year	P(yrs)	e	q	ω °	☊ °	i °
Kilston 1966 V	162000	0·9	2·385	154·5	155·4	40·3
Rudnicki 1967 II	–	1·0	0·420	79·7	75·0	9·1
Wild 1967 III	–	1·0	1·327	173·3	306·1	99·1
Seki 1967 IV	4420	0·9	0·457	144·4	199·8	106·5
Mitchell-Jones-Gerber 1967 VII	–	1·0	0·178	78·9	31·6	56·7
Ikeya-Seki 1968 I	89400	0·9	1·697	70·9	254·6	129·3
Wild 1968 III	–	1·0	2·660	103·5	208·4	135·3
Tago-Honda-Yamamoto 1968 IV	–	1·0	0·680	50·4	232·4	102·2
Whitaker-Thomas 1968 V	–	1·0	1·234	353·9	254·0	61·8
Honda 1968 VI	–	1·0	1·100	282·8	252·6	128·0
Bally-Clayton 1968 VII	–	1·0	1·772	26·9	318·7	93·2
Thomas 1969 I	18400	0·9	3·316	82·6	15·4	45·2
Fujikawa 1969 VII	–	1·0	0·774	299·0	191·7	9·0
Tago-Sato-Kosaka 1969 IX	419000	0·9	0·473	267·8	101·0	75·8
Daido-Fujikawa 1970 I	–	1·0	0·066	266·6	29·9	100·2
Bennet 1970 II	1730	0·9	0·538	354·2	223·9	90·0
Kohoutek 1970 III	83100	0·9	1·719	123·5	301·1	86·3
Suzuki-Sato-Seki 1970 X	–	1·0	0·406	318·5	292·9	60·8
Abe 1970 XV	–	1·0	1·113	96·6	21·0	126·7

APPENDIX V

Examples of Comet Groups (similar orbital elements)

Comet	q	$\tilde{\omega} = \Omega + \omega$*	Ω	i
770	0·603 (AU)	2°·1	88°·9	120°·5
1337	0·828	2·3	93·0	139·5
1468	0·830	1·4	71·1	142·0
1787	0·349	7·8	106·9	131·7
1799 I	0·840	3·7	99·5	129·1
1845	0·905	91·3	336·7	46·9
1854 IV	0·799	94·4	324·5	40·9
1925 VII	1·566	81·0	334·6	49·3
1822 I	0·504	192·7	177·4	126·4
1864 I	0·626	188·9	175·0	135·0
1893 IV	0·812	187·5	174·9	129·8
1743 II	0·523	247·0	6·1	134·4
1808 II	0·608	252·6	24·2	140·7
1857 III	0·367	249·6	23·7	121·0
1857 V	0·563	250·1	15·0	123·9
1762	1·009	104·0	348·6	85·6
1877 III	1·009	102·9	346·1	77·2
1668	0·067	248·8	358·6	144·3
1843 I	0·006	278·7	1·3	144·3
1880 I	0·005	279·9	6·1	144·7
1882 II	0·008	276·4	346·0	142·0
1887 I	0·010	266·3	324·6	128·5
1945 VII	0·006	270·7	321·6	137·0
1965 VIII	0·007	278·2	346·25	141·9
1970 VI	0·008	275·0	336·32	139·07

* $\tilde{\omega} = (\Omega + \omega)$ = longitude of perihelion formerly used in orbital work.

APPENDIX VI

Magnitudes of Comets/Meteorites/Meteoroids

The brightness or luminosity of a comet/meteorite /meteoroid is expressed in terms of *stellar magnitude.**

The magnitude of a comet varies according to its distance from the Sun (heliocentric distance = r) and its distance from the Earth (geocentric distance $= \Delta$). If a comet were to shine like a planet, its magnitude would vary according to the 'r^2 law' i.e. its luminosity would vary inversely with the square of the distance from the Earth and from the Sun. However, a comet's luminosity is found in practice to follow a light variation closer to the 'r^4–r^6 laws'. Nevertheless, each comet is different, and many show marked variations in brightness which are difficult to account for by any law. As a generalization short-period comets tend towards the 'r^4 law' while long-period comets the 'r^6 law'.

Comet magnitudes are usually computed from the formulae:

$(r^4)\ m = m_0 + 5 \log \Delta + 10 \log r$

or/ $(r^6)\ m = m_0 + 5 \log \Delta + 15 \log r$

where

m = mag. of comet

m_0 = absolute mag. of comet which is the mag. reduced to a distance of 1 AU from Earth and Sun.

Δ = distance from Earth in astronomical units (AU).

r = distance from Sun in astronomical units (AU).

Note:

The observed magnitude of a comet is usually expressed in terms of its integrated total brightness in relation to stellar magnitude. Since a star is a point source and a comet usually is very diffuse, this presents a difficult problem to the observer,

* Stellar magnitudes are logarithmically based. For example, a magnitude 1 star ($1^{m}.0$) is 2·52 times brighter than a magnitude 2 star ($2^{m}.0$). Magnitudes *brighter* than $0^{m}.0$ are denoted by a negative sign. (e.g. magnitude of Sun = –26·8, Moon –12·7). The faintest stars seen with the naked eye are approximately 6^{m}.

but with bright comets the magnitude is usually obtained (visually) by resorting to the practice of using extra-focal (out of focus) images of nearby stars (expanded to the size of the comet) and then compared with the diffuse image of the comet. The most accurate method is achieved by using photometry in conjunction with photography. Some observers quote magnitudes for the concentrated nuclear region only, and these are more easily estimated by direct reference to nearby stars of known brightness.

APPENDIX VII

Table of Principal Meteorite (Impact) Sites

	Latitude	*Longitude*
Aouelloul, Mauritania	21° 15′ N	012° 41′ W
Barringer, Arizona	35° 02′ N	111° 01′ W
Bosumtwi, Ghana	06° 32′ N	001° 23′ W
Boxhole, N. Territory, Australia	22° 37′ S	135° 12′ E
Brent, Ontario, Canada	46° 05′ N	078° 29′ W
Campo del Cielo, El Chaco, Argentina	27° 28′ S	061° 30′ W
Carswell, Saskatchewan, Canada	58° 27′ N	109° 30′ W
Charlevoix, Quebec, Canada	47° 32′ N	070° 18′ W
Clearwater, Quebec, Canada	56° 10′ N	074° 20′ W
Crooked Creek, Missouri	37° 50′ N	091° 23′ W
Dalgaranga, Western Australia	27° 45′ S	117° 05′ E
Decaturville, Missouri	37° 54′ N	092° 43′ W
Deep Bay, Saskatchewan, Canada	56° 24′ N	102° 59′ W
Dellen, Sweden	61° 50′ N	016° 45′ E
Flynn Creek, Tennessee	36° 16′ N	085° 37′ W
Gosses Bluff, N. Territory, Australia	23° 48′ S	132° 18′ E
Haviland, Kansas	37° 37′ N	099° 05′ W
Henbury, N. Territory, Australia	24° 34′ S	133° 10′ E
Holleford, Ontario, Canada	44° 28′ N	076° 38′ W
Howell, Tennessee	35° 15′ N	086° 35′ W
Janisjarvi, Karelia, USSR	61° 58′ N	030° 55′ E
Jeptha Knob, Kentucky	38° 06′ N	085° 06′ W
Kaalijarv, Estonia, USSR	58° 24′ N	022° 40′ E
Kentland, Indiana	40° 45′ N	087° 24′ W
Kofels, Austria	47° 13′ N	010° 58′ E
Lac Couture, Quebec, Canada	60° 08′ N	075° 18′ W
Lappajarvi, Finland	63° 10′ N	023° 40′ E
Liverpool, N. Territory, Australia	12° 24′ S	134° 03′ E
Lonar, Maharashtra, India	19° 59′ N	076° 34′ E

Manicouagan, Quebec, Canada	51° 23′ N	068° 42′ W
Manson, Iowa	42° 35′ N	094° 31′ W
Middlesboro, Tennessee	36° 36′ N	083° 36′ W
Mien, Sweden	56° 25′ N	014° 55′ E
Mistastin, Newfoundland, Canada	55° 53′ N	063° 18′ W
Monturaqui, Chile	23° 54′ S	068° 18′ W
New Quebec, Quebec, Canada	61° 17′ N	073° 40′ W
Nicholson, NWT, Canada	62° 40′ N	102° 41′ W
Odessa, Texas	31° 48′ N	102° 30′ W
Pilot, NWT, Canada	60° 17′ N	111° 01′ W
Pretoria Salt Pan, South Africa	25° 30′ S	028° 00′ E
Ries, Germany	48° 53′ N	010° 37′ E
Rochechouart, France	45° 50′ N	000° 56′ E
Roter Kamm, South West Africa	27° 45′ S	016° 17′ E
St Martin, Manitoba, Canada	51° 47′ N	098° 33′ W
Serpent Mound, Ohio	39° 02′ N	083° 25′ W
Sierra Madera, Texas	30° 36′ N	102° 55′ W
Sikhote-Alin, Siberia, USSR	46° 09′ N	134° 40′ E
Siljan, Sweden	61° 05′ N	015° 00′ E
Skeleton, Ontario, Canada	45° 15′ N	079° 27′ W
Steen River, Alberta, Canada	59° 31′ N	117° 38′ W
Steinheim, Germany	48° 02′ N	010° 04′ E
Strangways, N. Territory, Australia	15° 12′ S	133° 35′ E
Sudbury, Ontario, Canada	46° 36′ N	081° 11′ W
Talemzane, Algeria	33° 18′ N	004° 06′ E
Temimichat, Mauritania	24° 15′ N	009° 39′ W
Tenoumer, Mauritania	22° 55′ N	010° 24′ W
Vredefort, South Africa	27° 00′ S	027° 30′ E
Wabar, Saudi Arabia	21° 30′ N	050° 28′ E
Wahnapitae, Ontario, Canada	46° 44′ N	080° 44′ W
Wells Creek, Tennessee	36° 23′ N	087° 40′ W
West Hawk, Manitoba, Canada	49° 46′ N	095° 11′ W
Wolf Creek, Western Australia	19° 18′ S	127° 47′ E

APPENDIX VIII

The Major Annual Meteor Showers

Night Showers:	*Date of Peak Activity*	*Radiant Coordinates*		*Duration of Dectectable Meteors*	*Duration of Peak Days*	*Expected hourly rates*
		R.A.	*Dec.*			
Quadrantids	Jan. 3	231°	+50°	1–4 Jan.	0·5	50
Corona Australids	Mar. 16	245	–48	14–18 Mar.	5	5
Virginids	Mar. 20	190	00	Mar. 5–Apr. 2	20	5
Lyrids	Apr. 21	272	+32	19–24 Apr.	2	10
Eta Aquarids	May 4	336	00	Apr. 21–May 12	10	20
Ophuichids	June 20	260	–20	17–26 June	10	20
Capricornids	July 25	315	–15	July 10–Aug. 5	20	20
Southern Delta Aquarids	July 29	339	–17	July 21–Aug. 15	15	20
Northern Delta Aquarids	July 29	339	00	July 15–Aug. 18	20	10
Pisces Australids	July 30	340	–30	July 15–Aug. 20	20	20
Perseids	Aug. 12	46	+58	July 25–Aug. 17	5	50
Kappa Cygnids	Aug. 20	290	+55	18–22 Aug.	3	5

Night Showers:	*Date of Peak Activity*	*Radiant Coordinates R.A.*	*Radiant Coordinates Dec*	*Duration of Detectable Meteors*	*Duration of Peak Days*	*Expected hourly rates*
Orionids	Oct. 21	95	+15	18–26 Oct.	5	20
Southern Taurids	Nov. 1	52	+14	Sept. 15–Dec. 15	45	5
Northern Taurids	Nov. 1	54	+21	Oct. 15–Dec. 1	30	5
Leonids	Nov. 17	152	+22	14–20 Nov.	4	varies
Phoenicids	Dec. 5	15	–55	Dec. 5	0·5	50
Geminids	Dec. 13	113	+32	7–15 Dec.	6	50
Ursids	Dec. 22	217	+80	17–24 Dec.	2	5
Daylight Showers:						
Arietids	June 7	45	+23	May 29–June 19	20	60
Zeta Perseids	June 9	62	+24	1–17	15	40
Beta Taurids	June 29	87	+20	June 24–July 5	10	20
Piscids*	May 7–13	26	+25			30
o–Cetids*	May 21	30	–3			20
54–Perseids*	June 25	68	+33			50
α–Orionids*	July 12	87	+11			50
ν Geminids*	July 12	98	+21			60
λ Geminids*	July 12	111	+15			32
θ Aurigids*	July 25	87	+38			20

*Probably periodic showers rather than annual ones.

International Comet Symposium at Liège, 1965
Key to Plate 31, which faces page 145.

1 E. Roemer
2 Rh. Lüst
3 K. Wurm
4 N. Richter
5 F. L. Whipple
6 J. L. Greenstein
7 G. van Biesbroeck
8 P. Swings (host)
9 E. J. Opik
10 L. Biermann
11 G. Guigay
12 J. C. Brandt
13 Author
14 F. Dossin
15 D. McNally
16 L. Kresak
17 J. Rahe
18 V. Vanysek
19 O. Namba
20 Z. Sekanina
21 A. H. Delsemme
22 C. Arpigny
23 R. B. Southwater
24 W. Liller
25 D. Malaise
26 G. Righini

NAME INDEX

SUBJECT INDEX